T0156283

Lecture Notes in Business Information Processing 509

LNBIP reports state-of-the-art results in areas related to business information systems and industrial application software development – timely, at a high level, and in both printed and electronic form.

The type of material published includes

- Proceedings (published in time for the respective event)
- Postproceedings (consisting of thoroughly revised and/or extended final papers)
- Other edited monographs (such as, for example, project reports or invited volumes)
- Tutorials (coherently integrated collections of lectures given at advanced courses, seminars, schools, etc.)
- Award-winning or exceptional theses

LNBIP is abstracted/indexed in DBLP, EI and Scopus. LNBIP volumes are also submitted for the inclusion in ISI Proceedings.

Marta Campos Ferreira · Thomasz Wachowicz ·
Pascale Zaraté · Yu Maemura
Editors

Human-Centric Decision and Negotiation Support for Societal Transitions

24th International Conference
on Group Decision and Negotiation, GDN 2024
Porto, Portugal, June 2–5, 2024
Proceedings

Editors
Marta Campos Ferreira
University of Porto
Porto, Portugal

Pascale Zaraté
Toulouse University
Toulouse, France

Thomasz Wachowicz
University of Economics in Katowice
Katowice, Poland

Yu Maemura
University of Tokyo
Tokyo, Japan

ISSN 1865-1348 ISSN 1865-1356 (electronic)
Lecture Notes in Business Information Processing
ISBN 978-3-031-59372-7 ISBN 978-3-031-59373-4 (eBook)
https://doi.org/10.1007/978-3-031-59373-4

This Springer imprint is published by the registered company Springer Nature Switzerland AG
The registered company address is: Gewerbestrasse 11, 6330 Cham, Switzerland

Paper in this product is recyclable.

Preface

The field of Group Decision and Negotiation focuses on decision processes with at least two participants and a common goal. Such processes are complex and self-organizing, and constitute multi-participant, multi-criteria, ill-structured, dynamic, creative, and often evolutionary problems. Major approaches include:

- Information systems, in particular negotiation support systems (NSSs) and group decision support systems (GDSs);
- Cognitive and behavioral sciences as applied to group decision and negotiation;
- Conflict analysis and resolution;
- Applied game theory, experiment, and social choice;
- Artificial intelligence;
- Management science as it relates to group decision-making and negotiation.

Many research initiatives combine two or more of these approaches.

Group Decision and Negotiation can be performed in an intra-organizational as well as an inter-organizational context. Both consist of complex processes, including preference elicitation, proposals and counter-proposals, preference adjustment, and choice. Communication and decision-making are the two key steps in the Group Decision and Negotiation process and thus require sophisticated support in many ways.

Application areas of Group Decision and Negotiation include intra-organizational and inter-organizational coordination (as in operations management and integrated design, production, finance, marketing, and distribution functions, such as coordination of all phases of the life cycle of a product), computer-supported collaborative work and meetings, electronic negotiations, including negotiating agents and negotiation support systems, labor-management negotiations, inter-organizational, intercultural negotiations, environmental negotiations, and many others.

This book constitutes the refereed proceedings of the 24th International Conference on Group Decision and Negotiation, GDN 2024, which took place in Porto, Portugal, from June 2 to 5, 2024. It was organized in conjunction with the 10th International Conference on Decision Support System Technology under the overarching theme 'Human-Centric Decision and Negotiation Support for Societal Transitions.' This joint event took place at the Faculty of Engineering of the University of Porto.

At this joint conference, discussions were promoted on the human and technological aspects of decision-making processes to build bridges between the two domains:

1. From the human perspective, research should ensure that humans remain at the centre of the decisions, with participatory and negotiation processes that promote co-creation and co-design of technology, services, and regulations. Such reliable decision processes increase trust and fairness of the decisions.
2. From the technological perspective, research must demonstrate that technology can be trusted and that proposed solutions are safe, inclusive, and fair.

The 13 full papers presented in this volume were carefully single-blind reviewed by three reviewers each and selected from 100 submissions. They were organized in the following topical sections: Conflict Resolution; Preference Modeling for Group Decision and Negotiation; Collaborative and Responsible Negotiation Support Systems and Studies.

Organizing the conference and publishing the conference proceedings required the dedication of many members of the GDN community. We would like to express our gratitude to everyone who contributed and offered us their kind assistance in preparing for this event. Their efforts not only highlight the collaborative spirit of our community but also reflect the collective commitment to advancing the field of Group Decision and Negotiation. Particularly, we would like to thank:

- The Conference Chairs, Program Chairs, and Organizing Committee for the organizational and technical preparation of the conference, ensuring its smooth operation and effective management throughout the entire process.
- The Board and Members of the GDN INFORMS Community for their regular interaction and valuable advice, which greatly contributed to the planning and execution of the conference.
- The Stream Chairs, Programme Committee, and all Reviewers for their efforts in attracting compelling contributions to their tracks and providing informative and careful feedback to all submitted papers, thereby enhancing the quality and relevance of the conference program.
- The Authors of all 100 articles submitted to the conference, whose innovative ideas have allowed us to create a challenging and stimulating scientific event and compile a selection of the best works for this volume.
- The Springer Staff for their support in organizing this volume, helping us navigate the complexities of publication with ease and professionalism.

March 2024

Marta Campos Ferreira
Tomasz Wachowicz
Pascale Zaraté
Yu Maemura

Organization

Honorary Chairs

Adiel Teixeira de Almeida	Federal University of Pernambuco, Brazil
Fatima Dargam	SimTech-Reach Consulting, Austria

Conference Chairs

Sérgio Pedro Duarte	University of Porto, Portugal
Pascale Zaraté	University Toulouse Capitole - IRIT, France

Program Committee Chairs

António Lobo	University of Porto, Portugal
Boris Delibasic	University of Belgrade, Serbia
Tomasz Wachowicz	University of Economics in Katowice, Poland

Organizing Committee

Carlos Rodrigues	University of Porto, Portugal
Marta Campos Ferreira	INESC TEC, University of Porto, Portugal
Sara Ferreira	University of Porto, Portugal

Program Committee

Abdelkader Adla	University of Oran, Algeria
Alexis Tsoukias	Université Paris Dauphine-PSL, France
Ana Paula Costa	Federal University of Pernambuco, Brazil
Antonio de Nicola	ENEA, Italy
Ben C. K. Ngan	Worcester Polytechnic Institute, USA
Carolina Lino Martins	UFPE/UFMS, Brazil
Danielle Costa Morais	Federal University of Pernambuco, Brazil
Daouda Kamissoko	IMT Mines - Albi, France
Ewa Roszkowska	Bialystok University of Technology, Poland

François Pinet	INRAE, France
Fuad Aleskerov	HSE, Russia
Gert-Jan de Vreede	University of South Florida, USA
Haiyan Xu	Nanjing University of Aeronautics and Astronautics, China
Hannu Nurmi	University of Turku, Finland
Jason Papathanasiou	University of Macedonia, Greece
John Zeleznikow	Victoria University, Melbourne, Australia
Jorge Freire de Sousa	University of Porto, Portugal
José Moreno-Jiménez	Universidad de Zaragoza, Spain
José Pedro Tavares	University of Porto, Portugal
Konstantinos Vergidis	University of Macedonia, Greece
Liping Fang	Toronto Metropolitan University, Canada
Luís Dias	University of Coimbra, Portugal
Mareike Schoop	University of Hohenheim, Germany
María Teresa Escobar	Universidad de Zaragoza, Spain
Marta Campos Ferreira	University of Porto, Portugal
Masahide Horita	University of Tokyo, Japan
Michael Filzmoser	Vienna University of Technology, Austria
Muhammed-Fatih Kaya	University of Hohenheim, Germany
Nannan Wu	Nanjing University of Inf. Science and Technology, China
Nikolaos Matsatsinis	Technical University of Crete, Chania, Greece
Pavlos Delias	International Hellenic University, Greece
Przemysław Szufel	Warsaw School of Economics, Poland
Rosaldo Rossetti	University of Porto, Portugal
Rudolf Vetschera	University of Vienna, Austria
Sandro Radovanovic	University of Belgrade, Serbia
Sara Ferreira	University of Porto, Portugal
Sean Eom	Southeast Missouri State University, USA
Shawei He	Nanjing University of Aeronautics and Astronautics, China
ShiKui Wu	Lakehead University, Thunder Bay, Canada
Teresa Galvão	University of Porto, Portugal
Thiago Sobral	University of Porto, Portugal
Tomasz Szapiro	Warsaw School of Economics, Poland
Tung X. Bui	University of Hawaii, USA

Contents

Collaborative and Responsible Negotiation Support Systems and Studies

Conflict Resolution

An Analysis of the Predatory Fishing Conflict During the Piracema Period Through the Graph Model

Raí dos Santos Mota[1], Maísa Mendonça Silva[2], and Leandro Chaves Rêgo[1,3](✉)

[1] Graduate Program in Management Engineering,Universidade Federal de Pernambuco, Recife, PE 50670-901, Brazil
rai.santos@ufpe.br , leandro@dema.ufc.br
[2] Management Engineering Department, Universidade Federal de Pernambuco,Recife, PE 50670-901, Brazil
maisa.ufpe@yahoo.com.br
[3] Statistics and Applied Math Department, Universidade Federal do Ceará, Fortaleza, CE 60455-760, Brazil

Abstract. This paper presents an application of the graph model for conflict resolution aligned with the technique of option prioritizing in a conflict focused on the environmental context in Brazil. For the development of this study, two decision makers were considered, the fisherman and the government, with a total of four possible options to be taken by them. Moreover, for the fisherman, two different personality profiles (responsible and irresponsible) were considered in the conflict analysis. Particularly, this conflict portrays the issue of preservation of native species of aquatic environments during their reproductive period. In Brazil, this period is popularly known as the Piracema period, when the fisherman is supported by the public policy of the Unemployment Insurance for the Artisanal Fisherman. From the conflict analysis, some equilibrium states are suggested which highlight the importance of the intensification of inspection actions by the government.

Keywords: Graph Model for Conflict Resolution · Option Prioritizing · Unemployment Insurance for the Artisanal Fisherman

1 Introduction

According to the Food and Agriculture Organization of the United Nations - [5], small-scale fishing activities and aquaculture require attention due to their

The research was funded by Coordenação de Aperfeiçoamento de Pessoal de Nível Superior (CAPES) Brazil (financing code 001) and the second (grants 310729/2020-3 and 406697/2023-0) and third (grants 428325/2018-1, 308980/2021-2 and 406697/2023-0) authors were also funded by Conselho Nacional de Desenvolvimento Científico e Tecnológico (CNPq) Brazil.

M. Campos Ferreira et al. (Eds.): GDN 2024, LNBIP 509, pp. 3–15, 2024.
https://doi.org/10.1007/978-3-031-59373-4_1

fundamental importance for the livelihood of millions of families. They also contribute to the sustainability of aquatic ecosystems through responsible and efficient management.

In parallel to this, it is worth noting that these fishing activities are interrupted at different times of the year, which may vary according to each geographic region, in order to guarantee the adequate reproduction of aquatic species [11]. This time interval is known as the Piracema period in which the reproduction of these species takes place over four consecutive months, being of paramount importance to guarantee the subsistence of the fisherman.

However, this reproduction period known as Piracema is not always respected by the fisherman, whether due to lack of environmental awareness or poor socioeconomic conditions. For this reason, it is important to develop and comply with public policies that protect fish during this period. From this, the practice of predatory fishing arises, which is a type of illegal fishing, which is carried out in disagreement with the rules and regulations established by government agencies.

For the development of this study, the Graph Model for Conflict Resolution (GMCR) will be used, as it is a model easy to calibrate and analyze. In particular, the GMCR will be used for the elicitation of preferences through the technique of option prioritization, which is more usual in practical problems involving the GMCR. As can be seen in the studies by [19], this technique proves to be quite effective in eliciting the preferences of decision-makers in the face of strategic conflicts, whether they are unknown, uncertain or deterministic preferences.

For this study, a conflict with two DMs is considered, which are: the artisanal fisherman, who may or may not prefer predatory fishing, and the government, responsible for actions to catch predatory fishing, punish predatory fishermen and encourage fishermen environmentally responsible, depending on the scenario in which the fisherman is involved. For modeling the conflict, two fisherman behavior profiles are considered, responsible and irresponsible.

2 Background

2.1 GMCR

According to [12], the GMCR is a set of directed graphs, one for each of the decision makers (DMs) involved in the conflict, in which all have the same set of vertices that are the possible scenarios of the conflict, called feasible states. Furthermore, each of the DMs has a preference relation over the set of states. In mathematical terms, the GMCR consists of a set of DMs, defined by $N = \{1, 2, \ldots, n\}$, a set of feasible states $S = \{s_1, s_2, \ldots, s_m\}$, as well as a collection of directed graphs $D_i = \{S, A_i\}$ and a collection of binary preference relations on S defined by $\succ_i$, for each $i \in N$ [9,18].

In the GMCR, the preferences of DM i are modeled by an asymmetric binary relation ($\succ_i$) over S, where $s_1 \succ_i s_2$ means that DM i strictly prefers state s_1 to state s_2 [4,18]. In this sense, subsequent to this strict preference relation $\succ_i$, it is possible to derive two more preference relations on S: the weak preference

($\succcurlyeq_i$) and the indifference ($\sim_i$), where $s_1 \succcurlyeq_i s_2$ if DM i does not strictly prefer state s_2 to state s_1 and $s_1 \sim_i s_2$ if $s_1 \succcurlyeq_i s_2$ and $s_2 \succcurlyeq_i s_1$.

2.2 Stability Analysis of a Conflict

After modeling the conflict, a stability analysis is necessary, with the aim of establishing satisfactory solutions for the evaluated situations. To make this stability analysis possible in the GMCR, one must understand initially that the stability concepts reflect the strategies of the DMs in relation to their vision of the conflict and their perceptions of risk [8].

In the stability analysis of a conflict, the focal DM is used as a reference, which has two paths to choose from: the first, which is to remain with the conflict in its current state, and the second, which is to move the conflict to another state. However, if he/she decides to move the conflict to another state, the opposing DMs will also have the choice of whether or not to move the conflict. In this context, due to the fact that these decisions reflect the rational behavior of the DMs, when moving the conflict between the states, it appears that the DM can take into account the possible blockages that the other DMs can impose on their movements.

For each $i \in N$, and for each $s \in S$, we have the set of all states reachable by DM i, represented by $R_i(s)$, where the DM i can move unilaterally (in a single move), from state s, as formally expressed by [9,18]:

$$R_i(s) = \{s_1 \in S : (s, s_1) \in A_i\}.$$

Furthermore, it is worth describing the representation of the movements that lead to the unilateral improvement achieved by DM i. The set of all unilateral improvements for DM i, starting from state s is defined by:

$$R_i^+(s) = \{s_1 \in S : (s, s_1) \in A_i \text{ and } s_1 \succ_i s\}.$$

In general, for the study of stabilities, the possible movements that a DM can execute are evaluated, considering its actions in relation to a behavior based on the prediction and perception of risks related to the strategic conflict [8]. The stability concepts for conflicts with two DMs are recalled in Table 1, using DM i as the focal DM, where $N = \{1, 2\}$ represents the set of decision makers and $S = \{1, \ldots, m\}$ the set of feasible states.

Although the SSEQ concept is not implemented in the GMCR+ software [13], from the relations of this stability concept with the other 4 that were shown in [16], it is possible to determine the SSEQ stability of the states in the conflicts considered in this paper.

2.3 Option Prioritizing

The option prioritizing approach initially emerged from the preference tree concept proposed in [3,6]. This technique is based on eliciting preferences from the

Table 1. Stability concepts

Concept	Description
Nash [14]	DM i assumes that for any state to which it can move the conflict from a state considered Nash stable for it, that state will not be preferable to the initial state. **Definition 1.** *Let* $i \in N$, *state* $s \in S$ *is Nash stable for DM* i , *if and only if* $R_i^+(s) = \emptyset$.
GMR [10]	DM i evaluates its moves conservatively, since it believes that by unilaterally moving the conflict, its opponent, DM j, if possible, will react in a way that the conflict will end in a non-preferred state for DM i compared to the initial state s. **Definition 2.** *Let* $i \in N$, *state* $s \in S$ *is General Metarational stable for DM* i, *if and only if* $\forall s_1 \in R_i^+(s)$, *there exists a state* $s_2 \in R_j(s_1)$ *such that* $s \not\succ_i s_2$.
SMR [10]	DM i evaluates which move to execute, the reactions of its opponent, DM j, to the initial move, and finally, its own counter-reaction in response to its opponent's reaction; however, it is worth mentioning that there will be no states of greater preference in relation to the initial state s that can be reached by DM i, if s is SMR stable. **Definition 3.** *Let* $i \in N$, *state* $s \in S$ *is Symmetric Metarational stable for DM* i, *if and only if* $\forall s_1 \in R_i^+(s)$, *there exists a state* $s_2 \in R_j(s_1)$, *such that* $s \not\succ_i s_2$ *and* $s \not\succ_i s_3, \forall s_3 \in R_i(s_2)$.
SEQ [7]	DM i assumes that DM j moves the conflict not only thinking about sanctioning DM i's unilateral improvements, but also aiming for DM j's own improvements, unilaterally moving the conflict to a more preferred state for DM j, which is not preferred to the initial state for DM i. **Definition 4.** *Let* $i \in N$, *state* $s \in S$ *is sequentially stable for DM* i, *if and only if* $\forall s_1 \in R_i^+(s)$, *there exists a state* $s_2 \in R_j^+(s_1)$, *such that* $s \not\succ_i s_2$.
SSEQ [16]	DM i considers aspects of both SEQ and SMR stabilities in the sense that it assumes that DM j moves the conflict not only thinking about sanctioning DM i's unilateral improvements, but also aiming for DM j's own improvements, and DM i cannot escape from the sanction imposed by this move. **Definition 5.** *Let* $i \in N$, *state* $s \in S$ *is symmetric sequentially stable for DM* i, *if and only if* $\forall s_1 \in R_i^+(s)$, *there exists a state* $s_2 \in R_j^+(s_1)$, *such that* $s \not\succ_i s_2$ *and* $s \not\succ_i s_3, \forall s_3 \in R_i(s_2)$

DM, asking him/her to provide an ordered sequence of preference statements in descending order of priority from most important to least preferable, expressed in terms of Boolean combinations containing the conflict options [15].

In this sense, it is worth describing that the conflict options available to DM i are actions that may or may not be taken by this DM during the course of the conflict, and the set of these options can be denoted as $O_i = \{o_1^i, o_2^i, \ldots, o_{m_i}^i\}$. Furthermore, the set of all conflict options can be expressed as $O = \cup_{i=1}^{n} O_i$. Each combination of options taken by DMs in the conflict constitutes a possible conflict state. As not every combination of options may make sense in the context of a conflict, the set of feasible states S contains only those combinations that are feasible.

Every DM i, $i \in N$, provides an ordered sequence of preference statements $\phi_i = (\Omega_1, \Omega_2, \ldots, \Omega_{k_i})$ and each preference statement assumes either a value "True" (T) or "False" (F) in each conflict state. It is also worth noting that there are three types of preference statements: non-conditional, conditional or biconditional [15]. The non-conditional preference statement is defined as a combination of available options or numbers of options and the logical connectives, and may contain negation ("not" or -), conjunction ("and" or "&") and disjunction ("or" or "|"). The conditional or biconditional preference statements consist of two non-conditional statements linked by one of the connectives, the "IF", or the "if and only if", denoted by ("IFF"). In addition, parentheses ("(" and ")") are used in order to control and maintain the priority of operations in a preference statement.

An example of a preference statement is $(o_1^i \ \& - o_2^i) \ IF \ o_1^j$, which would be true in states where either the DM j does not take its first option or when DM j takes its first option and DM i takes its first option but does not take its second option. Thus, from the ordered list of preferences statements $\phi_i = (\Omega_1, \Omega_2, ..., \Omega_{k_i})$ of DM i in relation to the options of the conflict, its preferences on the states of the conflict are obtained, where those states that satisfy the first preference statements in the list are preferred. Formally, a state s of the conflict is strictly preferred to a state s' by the DM i if there exists some k^* such that $\Omega_k(s) = \Omega_k(s')$, for $k < k^*$, $\Omega_{k^*}(s) = T$ and $\Omega_{k^*}(s') = F$.

3 Application

3.1 Conflict Description

The conflict under study addresses an environmental context involving two DMs, the artisanal fisherman and the government. The focus of the study is on the preservation of native species of aquatic environments, that is, species that live in aquatic habitats and that need a certain period of time to carry out its reproduction, known as the Piracema period. This period has as main purpose to guarantee the sustainability of the species, since the predatory fishing carried out in periods of reproduction can endanger countless species of fish, crustaceans, among others.

In order to encourage respect for the Piracema period, Brazilian federal Law No. 8,287, of December 20, 1991, created the Unemployment Insurance Benefit for Artisanal Fishermen (SDPA), which is widely known as closed season insurance. However, according to the evaluation report of direct expenditures on the closed season insurance, it appears that this restriction of fishing in the period of Piracema, entails two social consequences: the weakening of traditional communities and the impossibility of the artisanal fisherman to obtain any income resulting from fishing during the ban period [1]. Therefore, these factors may cause the violation of the closed season period.

For the fisherman to receive SDPA payment, he/she must be registered in the General Registry of Fishing Activity (RGP). This is the record of the Federal Executive Branch whose purpose is to accredit individuals and legal entities to carry out fishing and aquaculture activities [1]. The public agency responsible for processing and qualifying the beneficiary, as well as operational monitoring of the benefits, is the National Institute of Social Security (INSS), but Caixa Econômica Federal is responsible for making the payment [1].

According to information from the National Treasury Secretariat (STN), portrayed in the expenditure report on the SDPA, it was found that from 1998 to 2018 there was a significant increase in expenditure on the SDPA, from R$13 million to R$2.54 billion [1]. It should be noted that this growth was due mainly to the increase in the number of beneficiaries. Therefore, it is clear that this public policy should carry out periodic surveys on the effectiveness of sustainability in the prohibited regions, as well as intensify the inspection actions related to both the release of the benefit and the preservation areas.

In this context, we observe the emergence of a conflict arising from the relationship between artisanal fishermen and the government, in which the behavior of one DM can affect the decisions of the other. In addition, for a better approach to this conflict, we used the GMCR+ software [13], taking into account two existing personalities of fishermen: the responsible fisherman, who values the preservation of species, and the irresponsible fisherman, that carries out predatory fishing even during the prohibition period. In this conflict, the fisherman has the unique option of performing predatory fishing (o_1^F) or not. The Government has three options: to catch (o_1^G) the action of predatory fishing, to punish (o_2^G) the fisherman who practices predatory fishing and to encourage (o_3^G) sustainability of species by providing a minimum wage for the fishermen who do not carry out predatory fishing. However, in practice, some states are not feasible and should be eliminated from conflict modeling. Thus, the states of the conflict in which the following options are taken simultaneously were not considered feasible: Fish, Catch and Encourage; Fish, Catch and Do Not Punish; Punish and Encourage; Do Not Catch and Punish; Catch and Do Not Punish; Do Not Fish and Punish, and Do Not Fish and Catch. The resulting feasible states are shown in Table 2.

As the last step of the modeling phase, we now seek to describe the results obtained from the application of the GMCR+ software, particularly with the use of the option prioritizing functionality present within the software. For this

Table 2. Feasible States of the Predatory Fishing Conflict

DM	Option	s_1	s_2	s_3	s_4	s_5
Fisherman	o_1^F (Fish)	N	Y	Y	N	Y
Government	o_1^G (Catch)	N	N	Y	N	N
	o_2^G (Punish)	N	N	Y	N	N
	o_3^G (Encourage)	N	N	N	Y	Y

use, it is required that the user initially provides an ordered list of preference statements in descending order of priority, denoted in terms of options and logical connectives. Subsequently, these statements are evaluated as true or false for each feasible conflict state.

The preferences analyzed in this conflict are about two DMs, the fisherman and the government. However, in relation to the fisherman, two different scenarios can be assumed: one in which he/she may behave responsibly during the closed season or another in which he/she is considered irresponsible and disregards any environmental awareness. The government, on the other hand, always has the same preference, in both contexts. In order to obtain the preference rankings used in this work, it was necessary to use the GMCR+ software, in which after filling in the information about the DMs' preference statements, it became possible to analyze the DMs' preference ranking on feasible conflict states. Next, preference rankings of these DMs and their specificities are shown.

We inform that the ordered sequences of preference statements used in this work were obtained after the exchange of point of views among the authors of this paper regarding the studied conflict, where the first author used his knowledge about ongoing similar conflicts in his local community. According to the Food and Agriculture Organization of the United Nations (FAO) and United Nations Development Program (UNDP) (2022), around 65% of artisanal fishermen in Brazil who receive closed season insurance have not completed elementary school, that is, they have up to 9 years of basic education. Based on this fact, it can be taken into account that their level of education is one of the main reasons that can influence their decisions, especially in relation to the prioritization of alternatives that do not involve environmental awareness. Lack of education also impairs the reasoning about the seriousness of their actions, such as, for example, the penalties resulting from carrying out predatory fishing and consequently being caught, in the context of this conflict.

Furthermore, according to FAO and UNDP (2022), approximately 50% of beneficiary fishermen did not report employment. Moreover, among those who declared some occupation, around 90% are self-employed or temporary workers in the rural sector. In view of this, it can be seen that the priority for incentives is something that is very preferable on the part of these DMs.

Regarding the government's preferences, it is observed that they are aligned with the objective of public policy. These preferences prioritize actions in which fishermen do not practice illegal fishing in exchange for incentives for their subsistence. Thus, the preferences of DMs in this conflict are based on environmental awareness and socioeconomic instability for subsistence.

According to the data reported by FAO UNDP (2022), as 65% of fishermen have a gap in relation to their basic education training, the other 35% completed middle or high school education. In this way, two profiles of fishermen may emerge, in which those with a higher level of education tend to prioritize preferences based on preserving the environment, while those with a lack of basic education will be more likely to commit predatory fishing.

Therefore, two factors emerge to be analyzed in this conflict, which are environmental awareness and incentive. Responsible fishermen prioritize receiving the incentive rather than the gains obtained from predatory fishing, while the irresponsible fishermen prioritize the return of predatory fishing that may be greater than the incentive, compromising stocks fishing grounds and leading to the extinction of species. Thus, it is clear that environmental awareness and the DM's livelihood in this period of prohibition go hand in hand during the piracema period.

In fact, according to a report carried out by the Federal University of Alagoas (UFAL) in partnership with the São Francisco River Basin Committee (CBHSF) in 2020, it was possible to observe little diversity and quantity of fish species in the Baixo São Francisco , where around 6 species represent 80% of the catches made in the river (UFAL, 2020). In this way, fishermen stand out as one of the key points to promote environmental preservation and awareness.

DMs' Preferences. DMs' lists of preference statements and the preference rankings derived from them are shown in Table 3.

We can observe that the government demonstrates a certain interest in state s_4, for which the fisherman does not execute the predatory fishing and will subsequently receive the incentive. On the other hand, state s_5 has a lower preference for the government, located in the last position in the ranking of priorities, since it refers to the state in which the fisherman performs predatory fishing and still receives the incentive.

Regarding the Responsible Fisherman, it is observed that this DM has state s_4 as a priority state, in which he/she does not carry out predatory fishing, consequently he/she will not be caught and not punished, and will later receive the incentive. State s_3, considered the least preferred for the responsible fisherman, is the state in which the fisherman practices illegal fishing and is caught and punished. It is worth noting that in this scenario, it appears that the profile of the responsible fisherman is more consistently aligned with the government's preferences, demonstrating greater effectiveness of public policy.

Finally, it can be seen that state s_5 is the most preferred for the Irresponsible Fisherman. This state is the one in which he/she performs predatory fishing and simultaneously receives the government incentive, without being caught or punished. On the other hand, state s_1 is the least preferable, and concerns the state in which the fisherman does not carry out predatory fishing and does not receive government assistance. Therefore, contrary to the previous scenario,

Table 3. DMs' Preference statement lists and rankings

Responsible Fisherman (RF)	Irresponsible Fisherman (IF)	Government (G)
$-o_1^F$	o_1^F	$-o_1^F$
$-o_1^G$	$-o_1^G$	o_1^F & o_1^G
o_3^G	$-o_2^G$	o_1^F & o_1^G & o_2^G
$-o_2^G$	o_3^G	$-o_1^F$ & o_3^G
$s_4 \succ_{RF} s_1 \succ_{RF} s_5 \succ_{RF} s_2 \succ_{RF} s_3$	$s_5 \succ_{IF} s_2 \succ_{IF} s_3 \succ_{IF} s_4 \succ_{IF} s_1$	$s_4 \succ_G s_1 \succ_G s_3 \succ_G s_2 \sim_G s_5$

in this context the fisherman has preferences that are very different from the government, prioritizing the practice of predatory fishing during the prohibition period.

3.2 Results and Discussions

In this section, we present the results of stability analyzes using the concepts presented in Subsect. 2.2.

Conflict Equilibria. In this study, the DSS GMCR+ was used for the analysis of equilibrium states, according to five definitions of conflict stability: Nash stability (R), general metarational stability (GMR), symmetric sequential stability (SMR), sequential stability (SEQ) and symmetric sequential stability (SSEQ). This analysis was made with the help of the DSS GMCR+ and also used relations among the SSEQ and the other four stability concepts to obtain its results. The equilibria states for the cases of Responsible and Irresponsible Fisherman are shown in Table 4.

For the conflict with the Responsible Fisherman, it is noted that states s_3 and s_4 are conflict equilibria under all stability concepts. It is even possible to state that the suggestions of these states as conflict equilibria are consistent with the preference rankings described previously for the Responsible Fisherman scenario, in which he/she does not carry out predatory fishing and will receive

Table 4. Conflict Equilibria - Responsible and Irresponsible Fisherman

Stability concept	Responsible					Irresponsible				
	s_1	s_2	s_3	s_4	s_5	s_1	s_2	s_3	s_4	s_5
Nash			X	X				X		
GMR	X		X	X				X		
SMR	X		X	X				X		
SEQ			X	X				X		
SSEQ			X	X				X		

the incentive, and if he/she does, consequently he/she will be caught, punished and will not get the benefit.

Finally, for the conflict with the irresponsible fisherman, it is noted that only state s_3 was suggested as an equilibrium of the conflict under all stability concepts. In this sense, it appears that, due to the fact that the Irresponsible Fisherman always opts for the action of fishing, this decision entails an involved consequence, that is, state s_3 demonstrates a scenario in which the fisherman takes the option of fishing, and consequently the government catches and applies the punishment, as well as does not transfer the incentive.

It is observed that state s_3 is an equilibrium in both cases, being an indication that this is a robust equilibrium that applies to both cases of fishermen. State s_3 represents the case in which the fisherman performs illegal fishing, is caught and punished. It is an equilibrium, because once the fisherman has been caught, he/she has no way to reverse the situation. However, this is a very undesirable equilibrium for the Responsible Fisherman and is unlikely to be chosen in this case. In the case of the Irresponsible Fisherman, to make other states become equilibria, the government must seek to generate incentives that modify the preferences of the Irresponsible Fisherman. For this, an approach like the one described in [17] can be useful and investigated in a future work.

The equilibrium states s_3 and s_4 reflect practical situations that may arise throughout this conflict. If state s_3 tends to become more frequent, then fish stocks will decrease, and the number of endangered species will increase, given that predatory fishing will intensify. To improve this situation, the government should seek to promote awareness training for these fishermen regarding the importance of respecting the breeding periods of species. Additionally, there is a need to strengthen monitoring and punitive actions against illegal activities. On the other hand if state s_4 becomes prevalent, then there will be a greater dissemination of sustainability and species preservation for future generations. In this scenario, fishermen will prioritize environmental protection during the spawning season. Simultaneously, they will receive governmental incentives to ensure their livelihoods. Therefore, for this scenario to become more frequent, financial investments from the government are required, both in terms of incentives and environmental education programs.

4 Conclusions

A predatory fishing conflict was analyzed through the application of the option prioritizing method within the GMCR with the help of the DSS GMCR+. It was possible to analyze the strategies and actions taken by each DM of the conflict, deepening in more details the preferences of the DMs, contributing to an improvement in the structuring of the conflict, and bringing more clarity to the relationship between artisanal fishermen and the government during the Piracema period. At the end, suggestions of states that favor more assertive decision-making are recommended.

The main contribution of modeling this conflict is that it becomes possible to observe that government actions can directly influence fisherman behavior and the effectiveness of SDPA public policy. Therefore, it appears that the options o_2 and o_4 related to flagrant and incentive are key pieces that should be better worked on by the government, mainly, in the intensification of inspection actions both for the release of SDPA and for the punishment of predatory fishing, in order to guarantee the sustainability of fishing stocks and the subsistence of fishermen during the Piracema period. This result is especially important because it highlights that the use of a conflict analysis approach as the one performed herein allows one to identify possible opportunities to better manage real world conflicts, for instance, by influencing DMs to change their behaviors.

It is worth mentioning that there are fishermen who practice predatory fishing even receiving the incentive, which is something totally illegal, a crime. On the other hand, there may be another majority of fishermen who are forced to practice this type of illegal fishing due to the delay in the transfer of the incentive by the responsible public agencies, since fishing is their only way of subsistence.

Thus, based on the analyzed results, it is suggested that the government seek initiatives to make the SDPA transfer process more efficient, so that it is released in a timely manner to the fisherman over the course of four months. In addition, the intensification of surveillance in regions during the prohibition period is essential to guarantee the preservation of species. It is mainly necessary to carry out a periodic survey of fish stocks, in particular the native species of each region; from these data, it becomes possible to ascertain which species require preservation and are in a period of extinction. Moreover, it would be possible to monitor the development of these species over time, thereby verifying whether the implemented public policy is being effective.

The modeling and analysis made in this work is similar to the work of Eggertsen et al. [2], which portrays the decrease in fish species that live in marine reefs, mainly caused by predatory fishing and a lack of satisfactory environmental control, causing the possible extinction of species, mainly due to artisanal fishing, which is responsible for the majority of catches. In this way, it is noted that the DMs, options and preferences vary according to the regions and the problem, since there are artisanal and industrial fishing, with the industrial one being on a larger scale, while the artisanal one is on a smaller scale. Therefore, there may be regions where there are two types of fishing in the same location, both artisanal and commercial, and consequently there will be a greater number of options due to the varied number of DMs. However, it is worth highlighting that preferences will be similar, as government bodies will prioritize preservation actions, while fishermen can take actions aimed at sustainability or not. Therefore, the conclusions obtained in this work tend to replicate in other locations where predatory fishing occurs.

References

1. Brazil, M.E.: Relatório de avaliação de gastos diretos: avaliação executiva do seguro defeso [online]. Technical report, Ministério da Economia, Brazil (2019)
2. Eggertsen, L., et al.: Complexities of reef fisheries in brazil: a retrospective and functional approach. Rev. Fish Biol. Fisheries **34**, 511–538 (2024)

3. Fang, L., Hipel, K., Kilgour, D.M., Peng, X.: A decision support system for interactive decision making - part I: model formulation. IEEE Trans. Syst. Man Cybern. Part C. App. Rev. **33**(1), 42–55 (2003)
4. Fang, L., Hipel, K.W., Kilgour, D.M.: Interactive Decision Making: the Graph Model for Conflict Resolution, vol. 3. John Wiley & Sons, New York (1993)
5. FAO: The State of World Fisheries and Aquaculture. FAO, Roma, Itália (2020)
6. Fraser, N.M., Hipel, K.W.: Decision support systems for conflict analysis. In: Singh, M.G., Hindi, K., Salassa, D. (eds.) Proceedings of the IMACS/IFORS 1st International Colloquium on Managerial Decision Support Systems, pp. 13–21 (1988)
7. Fraser, N.M., Hipel, K.W.: Solving complex conflicts. IEEE Trans. Syst. Man Cybern. Part C **9**(12), 805–816 (1979)
8. He, S., Hipel, K.W., Kilgour, D.M.: Analyzing market competition between airbus and Boeing using a duo hierarchical graph model for conflict resolution. J. Syst. Sci. Syst. Eng. **26**(6), 683–710 (2017)
9. Hipel, K., Kilgour, D., Fang, L.: The graph model for conflict resolution. In: Encyclopedia of Life Support Systems (EOLSS), Developed under the Auspices of the UNESCO, vol. 2, pp. 123-143. Eolss Publishers, Oxford, UK (2012). http://www.eolss.net/Sample-Chapters/C14/E1-40-04-01.pdf
10. Howard, N.: Paradoxes of Rationality: Games, Metagames, and Political Behavior. MIT Press, Cambridge (1971)
11. Ibama: Biodiversidade aquática: Períodos de defeso [online]. http://www.ibama.gov.br/biodiversidade-aquatica/periodos-de-defeso (2017). Accessed 14-Out-2022
12. Kilgour, D.M., Hipel, K.W., Fang, L.: The graph model for conflicts. Automatica **23**(1), 41–55 (1987)
13. Kinsara, R.A., Petersons, O., Hipel, K.W., Kilgour, D.M., Laurier, W.: Advanced decision support for the graph model for conflict resolution. J. Decis. Syst. **24**, 117–145 (2015). https://api.semanticscholar.org/CorpusID:31333046
14. Nash, J.F.: Equilibrium points in n-person games. Proc. Natl. Acad. Sci. U.S.A. **36**(1), 48–49 (1950)
15. Peng, X., Hipel, K.W., Kilgour, D.M., Fang, L.: Representing ordinal preferences in the decision support system GMCR II. IEEE Syst. Man Cybern. Comput. Cybernet. Simul. **1**, 809–814 (1997)
16. Rêgo, L.C., Vieira, G.I.A.: Symmetric sequential stability in the graph model for conflict resolution with multiple decision makers. Group Decis. Negot. **26**, 775–792 (2017)
17. Rêgo, L.C., Silva, H.V., Rodrigues, C.D.: Optimizing the cost of preference manipulation in the graph model for conflict resolution. Appl. Math. Comput. **392**, 125729 (2021)
18. Xu, H., Hipel, K.W., Kilgour, D.M., Fang, L.: Conflict Resolution Using the Graph Model: Strategic Interactions in Competition and Cooperation. SSDC, vol. 153. Springer, Cham (2018). https://doi.org/10.1007/978-3-319-77670-5
19. Yu, J., Hipel, K.W., Kilgour, D.M., Zhao, M.: Option prioritization for unknown preference. J. Syst. Sci. Syst. Eng. **25**(1), 39–61 (2016)

Uncertainty and Information Asymmetry in Underground Works: A Case Study

Muhammad Tajammal Khan(✉) and Masahide Horita

Department of Civil Engineering, The University of Tokyo, 7-3-1 Hongo, Bunkyo-Ku, Tokyo, Japan
{tajammal,horita}@g.ecc.u-tokyo.ac.jp

Abstract. Uncertainty, information asymmetry and inappropriate risk sharing negatively influence the delivery of construction projects and are primary causes of claims, disputes and mistrust between the contracting parties. The presence of uncertainty in subsurface soil conditions is the main challenge in construction of underground works. It leads to information asymmetry and opportunistic behavior by the contractor for acquiring additional benefits. The International Federation of Consulting Engineers (FIDIC) has introduced a Geotechnical Baseline Report (GBR) for underground works based on the principle of balanced risk sharing which may reduce the uncertainty in soil conditions. In this paper, principle-agent theory is used to understand the relationship of Employer (principle) and the Contractor (agent) and impact of uncertainty and information asymmetry on project delivery. Case study of an under-construction hydropower project by government organization is carried out to evaluate the contractor's opportunistic behavior for acquiring higher profits. It is concluded that uncertainty in soil conditions and asymmetric information led to the contractor's possible opportunistic behaviors.

Keywords: Soil Uncertainty · Asymmetric Information · Opportunistic Behavior

1 Introduction

Uncertainties in construction projects can arise from various sources including soil conditions and influence project cost, completion time and relationship between the stakeholders. Understanding the source of uncertainty is crucial to identify and allocate the potential risks between the concerned parties effectively. Uncertainties in the construction project cause information asymmetry between an employer and a contractor as the latter can collect more information about site conditions due to its construction expertise. Thus, the contractor may behave opportunistically to increase its profits, lowering the revenue. In economic literature, principal-agent theory is being used to investigate the impact of uncertainty and asymmetric information in a wide range of fields including construction projects [1–6].

To address information asymmetry, principals commonly employ monitoring strategies to observe agents' actions, aiming to reduce moral hazard and adverse selection

M. Campos Ferreira et al. (Eds.): GDN 2024, LNBIP 509, pp. 15–26, 2024.
https://doi.org/10.1007/978-3-031-59373-4_2

while safeguarding their interests [7]. This ensures agents act in the principal's interest within the principal-agent model [8]. Monitoring mechanisms, including formal methods like contracts and informal approaches, are utilized to tackle specific principal agent problems [9]. Besides contracts, formal mechanisms encompass policies, procedures, reporting structures, staffing, and training [10]. Direct observation by supervisors is a fundamental method and ensures compliance with agreed specifications. Incentive structures such as bonuses further align agent interests with those of the principal [11]. Despite the significant resource and time costs associated with monitoring, its effectiveness in mitigating agency problems and fostering trust is essential. Through promoting transparency, accountability, and performance incentives, monitoring mechanisms play a crucial role in aligning interests and maintaining the integrity of contractual relationships within the principal-agent framework [12].

Global construction activities are anticipated as USD 15.2 trillion between year 2020 to 2030 [13] and demand for underground infrastructure are increasing around the world including Japan for its geography and hilly terrain [14]. Underground projects are complex due to uncertainty in soil, challenging environment and early project completion demands. Uncertainty in soil conditions leads to delays and loss of resources as soil risks are unique in terms of anticipation and allocation between parties. Consequently, contracting parties face costly disputes, higher transaction costs and mistrust.

The International Federation of Consulting Engineers (FIDIC) has published first edition of "Conditions of Contract for Underground Works" which are also known as "FIDIC Emerald Book" wherein a new framework has been introduced for allocation of risks caused by uncertainty in subsurface soil between the employer and the contractor [15]. Under these conditions, baseline soil conditions are defined in the Geological Baseline Report (GBR) in detail, developing an agreed level of expected subsurface conditions. Simultaneously, the parties also acknowledge that the subsurface conditions cannot be determined precisely. Consequently, the risk of foreseeable soil conditions in GBR lies with the Contractor and the risk of unforeseeable conditions is allocated to the Employer. The Emerald Book is flexible towards the parties due to its inbuilt mechanism for adjustment in time and cost for completion of a project, as the baseline soil conditions and baseline production rates are also included in the contract.

This study mainly aims to examine the impact of uncertainty in soil conditions and asymmetric information on project outcome through a case study and elaborate the opportunistic behavior of the Contractor.

2 Literature Review

Construction projects are not only influenced by stakeholders' effort levels but also impacted by the uncertainties beyond control [16]. Transactions performed under uncertain conditions have contingencies compared to certain environments facilitating the contractor to reduce effort level to safeguard its interests [17]. Uncertainty enables the contractor to collect more information about the site after signing the contract and is better informed which is known as information asymmetry [18, 19]. Huge information asymmetry prevails between government employer and the contractor during construction stage [18] as employer can neither grasp true information nor fully understand construction methodologies [20]. Uncertainty and information asymmetry lead to transactions with substantial opportunistic behavior by the contractor [4, 18] and it is important for employer to establish an effective mechanism to reduce contractor's opportunistic behavior [17].

Contractor with private information may conceal true information under uncertainty and report wrong information to seek higher profits [4] through opportunistic behavior. Contractor's tendency to behave opportunistically instead of employer's interest is known as moral hazard in economics and influences the successful completion of a project [16]. Asymmetric information is the primary cause of loss of efficiency and source of risk in construction projects leading to opportunistic behavior [21, 22]. Opportunistic behavior leads towards delayed project completion, disputes, increased transaction cost and mistrust. Realizing information asymmetry could benefit the construction industry to enhance the construction efficiency and trust between project stakeholders [22]. Employer can increase project performance through investing higher care in uncertainty and asymmetric information cases [4]. Therefore, each party should pay huge attention to information collection and management for better collaboration in a construction project [21].

The concept of information asymmetry has been studied on construction projects for the last 20 years either through questionnaires or economic theory like principal-agent theory [22]. Researchers have used principal-agent theory and developed incentive mechanisms for risk sharing [16, 23], risk management [24], reducing opportunistic behavior [25], understanding double moral hazard [26], uncertainty and information asymmetry [4–6]. Contractors are more likely to hide true information and keep it private rather than public. True information can benefit the employer and shall provide information rent to contractor under asymmetric environment [4, 5]. It is also necessary to devise a mechanism for construction projects for mining the information from the better-informed party to reduce information asymmetry [19]. These incentive mechanisms are complex and contain constraints owing to the fact of uncertainty and asymmetric information [27].

Literature review on principal-agent model [4, 5] demonstrates that effort level, project completion time and employer's profit are negatively influenced by presence of uncertainty and information asymmetry. Contractor will reduce the effort level under uncertainty and information asymmetry impacting the optimum decisions for the employer. The model also establishes that employer can extract the entire profit from the contractor in case of symmetric information and will be willing to invest for acquiring the true information. The information value is always positive, showing that the employer will always acquire profits after knowing the true information about the site. Information value is directly related to uncertainty level and employer shall invest more to acquire true information under highly uncertain or volatile environment.

Main challenge in underground works is to deal with inherent soil uncertainty arising from variation in geological formations along the tunnel alignment [29–32]. Extensive investigations at design stage do not guarantee to anticipate the precise geology and parties face surprises during construction due to uncertainty in soil conditions [31, 32]. Excavation methodology, preliminary support system and lining type for the tunnels are decided based on encountered soil condition at site [29, 33]. It demands flexibility, balanced risk allocation [23] and collaborative decisions for project success [28, 29]. However, employers generally amend contract clauses transferring the soil risks to the contractor to avoid claims for additional time and/or payment [34], which is a kind of opportunistic behavior from the employer as well.

FIDIC Emerald Book is the first international form of contract designed explicitly for underground works catering the uncertainty in soil condition. It allocates foreseeable conditions to the contractor and unforeseeable conditions to the employer. Geotechnical Baseline Report (GBR) is incorporated in FIDIC Contract and aims for equitable and transparent allocation of subsurface risk between the parties. This is achieved through establishing the anticipated baseline subsurface conditions through revelation of complete soil information at the procurement stage. GBR becomes a common document providing the contractual basis for expected baseline subsurface conditions i.e. rock classes and baseline production rates. These baseline conditions are used for adjustment of contract price and/or completion time considering encountered soil conditions at site. GBR requires intradisciplinary coordination for deeper understanding of project and its technical and management challenges ahead of construction. Project delays, cost overrun and claims are mainly developed from poor definition and management of GBR during construction phase [33].

3 Case Study of Hydropower Project

Project data of an under-construction hydropower project comprising complex network of underground works is analyzed to understand the impact of uncertainty in soil conditions and asymmetric information on project performance. Project consists of 200 m high dam having installed capacity of about 800 MW with estimated construction cost of approximately USD 2,500 Million. Underground works consist of access tunnels, diversion tunnels and power tunnels which are at construction stage. The layout plan of underground works developed in the Autodesk Revit (2024), a Building Information Modelling (BIM) software, is shown in Fig. 1.

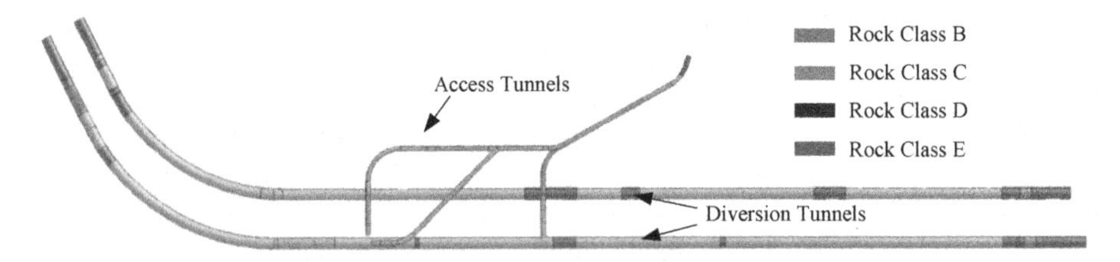

Fig. 1. Layout Plan of Access Tunnels and Diversion Tunnels

FIDIC-MDB Harmonized Edition (2010), also known as "pink book" is used to define the conditions of the Contract between the employer (government) and the international contractor. Extensive soil investigations were conducted on site during the design stage. However, FIDIC proposed GBR defining the anticipated baseline rock classes along the tunnel alignment was not envisaged, demonstrating huge uncertainty in ground conditions. The employer has also transferred all the foreseeable and unforeseeable soil risks to the contractor to avoid future claims for additional time for completion and/or cost caused by any variation in soil conditions, showing opportunistic behavior on its part. These amendments in the contract are against the spirit of balanced risk sharing approach and collaborative decision making as outlined in the FIDIC proposed GBR and may become a source of future claims and disputes between the parties.

Six (6) types of rock classes are defined in the Contract Documents to differentiate the effort level and advance exaction rate in view of uncertainty in soil conditions along the tunnels. Rock class is determined using a mechanism jointly developed by the contractor and the engineers for Q-System [35] before each blast to decide the excavation diameter, excavation length and primary support requirements. Table 1 outlines the defined rock classes, respective excavation length and approximate time for support installation from the project data.

Table 1 shows that excavation and preliminary support installation can be completed in short time for better rock class. Actual construction progress record depicts that contractor has influenced the engineer to allow increased excavation length for each rock class to shorten the construction period with lower effort indicating its tendency to utilize its private information about the site.

Table 1. Excavation, Support and Linning Classes for Diversion Tunnels

Rock Class	Description	Excavation Radius (m)	Excavation Length (m)	Approx. Time for Support Installation
A	Very Good	8.325	2.4*	-
B	Good	8.375	2.0 (3.5**)	-
C	Fair	8.450	1.6 (3.0)	20 ~ 30 h**
D	Poor	8.525	1.2 (2.4)	38 ~ 48 h
E	Very Poor	8.600	0.8 (1.6)	48 ~ 65 h
F	Extra Ordinary Poor Ground Tunnel ***	-	-	-

Notes:
** Allowed in construction drawings*
*** Estimated from actual site data / progress reports*
**** Excavation of tunnel portals on either end, excavation under passing nullah shall be classified as Class F for special rock support treatment*

Overall completion time for tunnels can neither be extended for poor rock class nor shortened for better rock class under the signed conditions of contract as the risk for any variation in soil strata is allocated to the contractor. Hence, uncertainty in soil condition and private information about site may motive the contractor to behave opportunistically and influence the engineer to determine better rock class to adjust delays on its part and avoid any plenty for delayed completion.

Next section outlines the results after analysis of progress data along access tunnels and diversion tunnels at the project site.

3.1 Access Tunnels

Round arch type tunnels (6.4 m diameter) are constructed at the first stage to provide access at the middle of diversion tunnels for additional construction faces to increase pace of construction. The top level of access tunnel varies between 370 masl and 405 masl with similar rock class properties described in Table 1. Figure 2 shows the unit rate to excavate each unit of rock class and comparison of excavation quantities in contract (BOQ), actual excavation quantities (ACT) and from BIM virtual model (BIM). Variation in the unit rate shows that it is beneficial for the contractor, if agreed rock class is better bidding higher rates for excavation.

Comparing the ACT with the BOQ demonstrates a shift from poor rock class (D, E and F) towards better rock class (B and C). Determining better rock class provides higher unit rate for excavation and leads to early completion with lower effort level as contractor can increase excavation length and install primary support in short time.

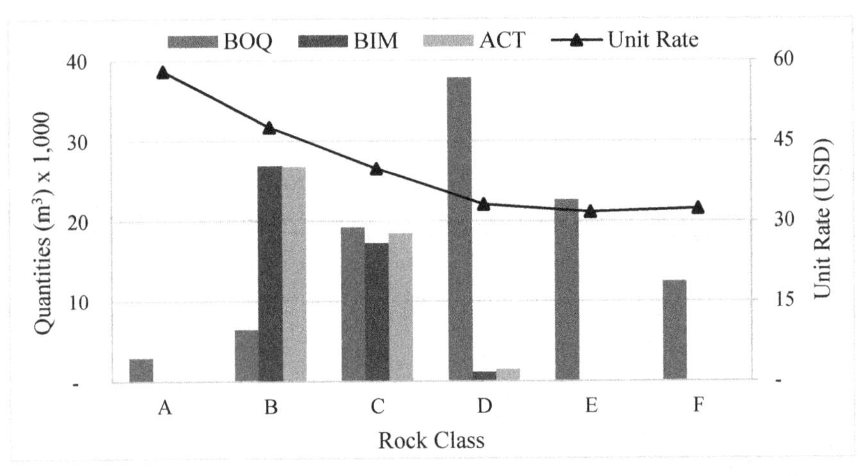

Fig. 2. Comparison of unit price and excavation quantities for Access Tunnel

3.2 Diversion Tunnels

Two parallel tunnels of 15 m finished diameter are being constructed and top level varies between the elevation 360 masl to 373 masl, almost same as of access tunnels. A comparison of unit price and excavation quantities for each rock class measured for the diversion tunnels is plotted in Fig. 3.

The plot shows a shift of excavation quantities from poor rock towards a better class like the access tunnels. It is an indication of contractor's opportunistic behavior to acquire benefits in terms of additional payment and shortened time for construction.

In continuation, Fig. 1 shows that rock has been determined as "class D" at the end portals (portion) of diversion tunnels while ignoring the classification outlined in the contract as "class F" (Table 1 Note) based on site investigation at design stage. Class

D requires less support and can be installed in a shorter time with less effort. It may have compromised the safety of the tunnels by providing week primary support system. Diversion tunnels were facing large settlement at end portals of each tunnel exceeding the permissible limits especially for class D and class E. The contractor was instructed to provide additional support in those areas to stabilize the tunnels, which provided a basis for future claims.

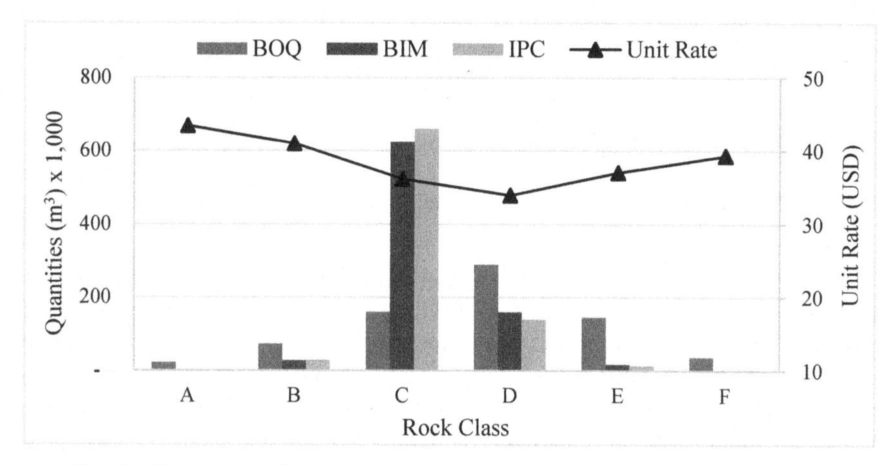

Fig. 3. Comparison of unit price and excavation quantities for diversion tunnels

To further explore the possible tendency of contractor's opportunistic behavior, the determined rock classification at the junction points of access tunnels and diversion tunnels are plotted in Autodesk Revit as shown in Fig. 4. Technically the interaction points have the same soil conditions, but different rock classification is observed at construction stage. It gives the impression that better class, mostly class B, has been determined along access tunnels benefiting the contractor in terms of additional payments and early completion with lower effort.

3.3 Contractor's Claims and Disputes

Comparison of excavation quantities depicts determination of better rock classes along the tunnels showing favorable conditions at site. However, the contractor has submitted the claims for extension of time and/or additional payments highlighting weak geological conditions and supplementary preliminary rock support indicating that the contractor is making efforts to acquire additional benefits, which is also a kind of opportunistic behavior. A summary of selected contractor's claims related to diversion tunnels and subsurface conditions is provided in Table 2.

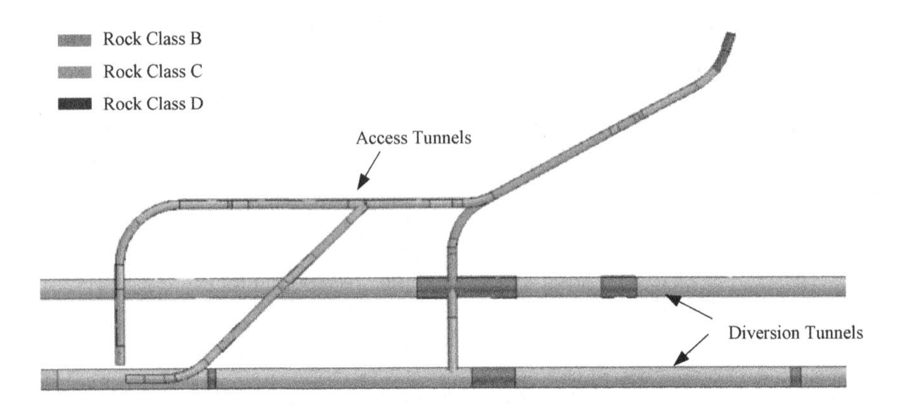

Fig. 4. Comparison of Rock Class at the intersection of access tunnels and diversion tunnels

The contract assigns all foreseeable and unforeseeable subsurface soil risks to the Contractor without any provision of additional time claim. Since the Contractor is uncertain about the approval of its claims under the provisions of Contract, disputes understandably arouse between the parties and created an environment of mistrust. Consequently, the parties are spending more time on data collection, analysis and claims rather than focusing on the original construction activities.

Table 2. Claim for extension of time for completion of diversion tunnels

S/No	Description	Claim (days)
1	Weak geological conditions	325
2	Increase in preliminary rock support	299
3	Variation in scope	30
4	Increase in length of diversion tunnels	140

In summary, the principal's (employer) goal is early project completion avoiding future claims through transferring the subsurface risks to the agent (contractor). However, the agent's (contractor) interest is to increase the profits from the project. Therefore, both the principal and agent have the conflicting goals and each party is pursuing its interests through an opportunistic behavior influencing their commitment to the project success.

Considering the aforesaid scenarios pointing the contractor's opportunistic behavior, incorporating FIDIC Emerald Book GBR in the Contract may theoretically benefit the employer in the following three ways:

- Firstly, inclusion of baseline conditions in the form of rock classes and respective baseline production rates in the contract may prompt to determine rock class carefully as deciding better class will shorten the construction time of the project. This situation will provide a trade-off between increased cost and shortened time for the contractor reducing opportunistic behavior.

- Secondly, excavation quantities (ACT) are reduced by approximately 50% for access tunnel compared to the BOQ, showing reduction in scope of work which provides favorable conditions compared to baseline quantities in the contract. Therefore, reduced quantities and/or scope will automatically adjust the completion time towards early completion and vice versa the contractor's claims for increase in diversion tunnel length and/or scope of work towards delayed completion. This balanced adjustment approach may avoid the disputes which are being face on the project.
- Thirdly, GBR proposed baseline rock classification along the tunnel alignment based on soil investigations at design stage will help to reduce the uncertainty in soil conditions. It will encourage the site engineers to consult design and site data to reach a better decision for rock class. Rock classification provided in the contract for the tunnel end points "class F" of this project is a sign regarding importance of baseline classification in GBR. Any deviation of encountered rock class from baseline GBR will demand extra care regarding determination procedure adopted for rock classification.

However, further research is necessary to understand the possible functions of GBR, especially its role to minimize the likely opportunistic behavior from the contractor.

4 Conclusion

This paper has provided factual evidence that hints the opportunistic behavior suggested theoretically by the literature. The employer transferred the risks due to uncertainty in subsurface conditions to the contractor adopting conservative approach and restricted the contractor's right to claim additional time for encountering the onerous conditions. At the same time, the employer did not establish the baseline subsurface conditions at the procurement stage showing huge uncertainty at site for both the parties. The employer cannot monitor the contractor's effort to complete excavation and preliminary rock support for a particular rock class. Therefore, the contractor had private information about the site and its level of effort causing information asymmetry. Uncertainty in soil and information asymmetry may have motivated the contractor to show opportunistic behavior to determine better rock class influencing the benefits to the employer and safety of tunnels. It has aroused the variations, disputes, and mistrust and has increased the transaction cost leading to loss of efficiency. Therefore, the uncertainty and information symmetry have become one of the main causes of project delay.

This study also highlights that the employer shall realize the importance of information asymmetry as it can enhance project performance for the employer. The employer shall invest in collecting and analyze the information especially under higher uncertainty and information asymmetry cases. It is therefore proposed that the employer shall devise a monitoring mechanism for construction projects rather providing information rent to the contractor as proposed in economic literature. BIM model can be developed to anticipate and visualize the soil conditions along the alignment of tunnel and can share information to all the parties for collaborative decision-making process.

5 Limitations

This case study has considered specific variation in unit rates for excavation in each class and without any baseline rock classes or baseline GBR. It will be interesting to analyze the projects having baseline GBR and different combination of variation in unit rates for rock classes.

6 Future Research Work

The goal of this research is to propose a monitoring framework that can improve the prediction certainty about variation in rock classes in the GBR and develop a mechanism for the employer to reduce opportunistic behavior, transaction cost and increase the trust and collaboration decision-making process for improving the project performance.

Acknowledgments. This work was supported by JSPS KAKENHI Grant Number JP22H01561.

Disclosure of Interests. The authors have no competing interests to declare that are relevant to the content of this article.

References

1. Bergmann, R., Friedl, G.: Controlling innovative projects with moral hazard and asymmetric information. Res. Policy **37**, 1504–1514 (2008). https://doi.org/10.1016/j.respol.2008.05.004
2. Lee, M., Su, L.: Study on the asymmetry information problem based on principal –agent theory. J. Bus. Econ. Manage. **4**(2), 040–045 (2016). https://doi.org/10.15413/jbem.2016.0110
3. Shi, S., Yin, Y., An, Q., Chen, K.: Optimal build-operate-transfer road contracts under information asymmetry and uncertainty. Transp. Res. Part B **152**, 65–86 (2021)
4. Yang, K., Zhao, R., Lan, Y.: Impacts of uncertain project duration and asymmetric risk sensitivity information in project management. Int. Trans. Oper. Res. **23**, 749–774 (2016). https://doi.org/10.1111/itor.12156
5. Qi, L.: Project duration contract design problem under uncertain information. Soft. Comput. **22**(17), 5593–5602 (2017). https://doi.org/10.1007/s00500-017-2527-5
6. Yao, M., Wang, F., Chen, Z., Ye, H.: Optimal incentive contract with asymmetric cost information. J. Constr. Eng. Manage. **146**(6), 04020054 (2020). https://doi.org/10.1061/(ASCE)CO.1943-7862.0001832
7. Ross, S.A.: The economic theory of agency: the principal's problem. Am. Econ. Rev. **63**(2), 134–139 (1973)
8. Holmström, B.: Moral hazard and observability. Bell J. Econ. **10**(1), 74–91 (1979)
9. Langerfield-Smith, K., Smith, D.: Management control systems and trust in outsourcing relationships. Manag. Account. Res. **14**(3), 281–307 (2003)
10. Badenfelt, U.: I trust you, I trust you not: a longitudinal study of control mechanisms in incentive contracts. Constr. Manag. Econ. **28**(3), 301–310 (2010)
11. Baker, G., Gibbons, R., Murphy, K.J.: Relational contracts and the theory of the firm. Quart. J. Econ. **117**(1), 39–84 (2002)

12. Tirole, J.: Hierarchies and bureaucracies: on the role of collusion in organizations. J. Law Econ. Organ. **2**(2), 181–214 (1986)
13. Robinson, G., Leonard, J., Whittington, T.: Future of Construction: A Global Forecast for Construction to 2030; Marsh & Guy Carpenter, Oxford Economics: London, UK (2021)
14. Hata, K.: Evaluation of tunnel rock mass using deep learning. Jpn. Soc. Civil Eng. **10**, 260–274 (2022)
15. FIDIC Conditions of Contract for Underground Works (2019): Emerald Book; first edition
16. Hosseinian, S.M., Carmichael, D.G.: Optimal incentive contract with risk-neutral contractor. J. Constr. Eng. Manag. **139**(8), 899–909 (2013). https://doi.org/10.1061/(ASCE)CO.1943-7862.0000663
17. You, J., Chen, Y., Wang, W., Shi, C.: Uncertainty, opportunistic behavior, and governance in construction projects. The efficacy of contracts. Int. J. Project Manage. **36**, 795–807 (2018). https://doi.org/10.1016/j.ijproman.2018.03.002
18. Kraus, S.: An overview of incentive contracting. Artif. Intell. **83**, 297–346 (1996)
19. Ceric, A.: Strategies for minimizing information asymmetries in construction projects: project managers' perceptions. J. Bus. Econ. Manag. **15**(3), 424–440 (2014). https://doi.org/10.3846/16111699.2012.720601
20. Liu, J., Wang, Z., Skitmore, M., Yan, L.: How contractor behavior affects engineering project value-added performance. J. Manage. Eng. **35**(4), 04019012 (2019). https://doi.org/10.1061/(ASCE)ME.1943-5479.0000695
21. Xiang, P., Huo, X., Shen, L.: Research on the phenomenon of asymmetric information in construction projects: the case of China. Int. J. Project Manage. **33**, 589–598 (2015). https://doi.org/10.1016/j.ijproman.2014.10.007
22. Ceric, A., Ivic, I.: Risks caused by information asymmetry in construction projects: a systematic literature review. Sustainability **15**, 9979 (2023). https://doi.org/10.3390/su15139979
23. Chang, C.Y.: Principal-agent model of risk allocation in construction contracts and its critique. J. Constr. Eng. Manage. **140**(1), 04013032 (2014). https://doi.org/10.1061/(ASCE)CO.1943-7862.0000779
24. Zhu, J., Hertogh, M., Zhang, J., Shi, Q., Sheng, Z.: Incentive mechanisms in mega project-risk management considering owner and insurance company as principals. J. Constr. Eng. Manage. **146**(10), 04020120 (2020). https://doi.org/10.1061/(ASCE)CO.1943-7862.0001915
25. Liu, J., Gao, R., Cheah, C.Y.J., Luo, J.: Incentive mechanism for inhibiting investors' opportunistic behavior in PPP projects. Int. J. Project Manage. **34**, 1102–1111 (2016). https://doi.org/10.1016/j.ijproman.2016.05.013
26. Shi, L., He, Y., Onishi, M., Kobayashi, K.: Double moral hazard and risk-sharing in construction projects. IEEE Trans. Eng. Manage. **68**(6), 1919–1929 (2021). https://doi.org/10.1109/TEM.2019.2938261
27. Hart, O.: Incomplete contracts and control. Am. Econ. Rev. **107**(7), 1731–1752 (2017). https://doi.org/10.1257/aer.107.7.1731
28. Mitelman, A., Gurevich, U.: Implementing BIM for conventional tunnels - a proposed methodology and case study. J. Inf. Technol. Constr. **26**, 643–656 (2021). https://doi.org/10.36680/j.itcon.2021.034
29. Rich, F., Giai Via, C., Bitetti, B., Ragazzo, G., Pepiot, J., Lione, S.: Tunnel Euralpin Lyon-Turin CO08 – BIM implementation in conventional tunneling. In: Anagnostou, G., Benardos, A., Marinos, V.P. (eds.) Expanding Underground - Knowledge and Passion to Make a Positive Impact on the World: Proceedings of the ITA-AITES World Tunnel Congress 2023 (WTC 2023), 12-18 May 2023, Athens, Greece, pp. 2869–2876. CRC Press, London (2023). https://doi.org/10.1201/9781003348030-346

30. Fabozzi, S., Biancardo, S.A., Veropalumbo, R., Bilotta, E.: I-BIM based approach for geotechnical and numerical modelling of a conventional tunnel excavation. Tunn. Undergr. Space Technol. **108**, 103723 (2021). https://doi.org/10.1016/j.tust.2020.103723
31. Sibaii, M.E., Granja, J., Bidarra, L., Azenha, M.: Towards efficient BIM use of geotechnical data from geotechnical investigations. J. Inf. Technol. Constr. (ITcon) **27**, 393–415 (2022). https://doi.org/10.36680/j.itcon.2022.019
32. Erharter, G.H., Weil, J., Bacher, L., Heil, F., Kompolschek, P.: Building information modelling-based ground modelling for tunnel projects. Tunn. Undergr. Space Technol. **135**, 105039 (2023). https://doi.org/10.1016/j.tust.2023.105039
33. Gomes, A.R.A.: Considerations on the practical development of the geotechnical baseline report (GBR) for the FIDIC emerald book and similar contract forms. In: ITA-AITES World Tunnel Congress, WTC2020 and 46th General Assembly Kuala Lumpur Convention Centre, Malaysia 15–21 May 2020
34. Construction Law International, Volume 17, Issue, 2 June 2022 (A Committee publication from the IBA Energy, Environment, Natural Resources and Infrastructure Law Section tinyurl.com/IBA-SEERIL)
35. Using the Q-system, Rock mass classification and support design: Handbook; New Edition: Norweign Geotechnical Institute, Postboks 3930, Ullevål Stadion, 0806 OSLO, Norway (2022). www.ngi.no
36. Hydropower Project Documents and Data: Detailed Engineering Design, Bidding Document, Contract Document, Progress Reports, and other related data

Developing a Multi-phase Stakeholder Game Framework for Recyclable Resource Management System

Jing Ma(✉), Dongbin Wang, Haimei Li, and Zhengbing Guo

School of Public Policy and Administration, Xi'an Jiaotong University, No. 28 Xianning Road West, Xi'an 710049, China
jing.ma@xjtu.edu.cn, {dongbin.wang,2625346624, gzb7965}@stu.xjtu.edu.cn

Abstract. Understanding the game relationships and behaviors of stakeholders is crucial for effectively promoting the management of recyclable resources. Aiming at the problems existing in the stakeholder game models in recyclable resource management, this study proposed a novel multi-phase stakeholder game framework for the recyclable resource management system, which fully takes into account stakeholders at all levels in the life cycle of recyclable resource management. In order to concretely analyze the game framework, cross-sectional and longitudinal analyses of the game relationships among and within stakeholder groups were conducted. By providing a more comprehensive understanding and explanation of stakeholder game relationships and behaviors in recyclable resource management, the proposed game framework could better guide the overall and various levels of recyclable resource management practices.

Keywords: Recyclable Resource Management · Decision Support Systems · Game Theory · Conflict Resolution

1 Introduction

Recyclable resources are defined as valuable waste that can be recirculated in the market after being collected, processed [1], and remanufactured, and are regarded worldwide as a pathway to achieving a circular economy and promoting sustainable development [2]. Despite the promising future of recyclable resources, recycling remains a global challenge. This is particularly evident in developing countries that have poor resource management, less awareness of recycling, and the need to deal with large amounts of imported waste. Wang et al. and Zhang et al. stated that the main reason for the low recycling rates is a lack of stakeholder motivation [3, 4], evidence suggests that the effectiveness of waste recycling policies significantly influences the level of waste recycling [5]. Improving recycling rate has far-reaching significance for the recycling resource management of developing countries [6], and even global sustainable development. The key to the enthusiasm of stakeholders is to satisfy their interests and coordinate the

M. Campos Ferreira et al. (Eds.): GDN 2024, LNBIP 509, pp. 27–37, 2024.
https://doi.org/10.1007/978-3-031-59373-4_3

relationships among them [7, 8]. Effective and efficient MSW management is widely recognized as a crucial factor for future social development. This involves not just technical innovation but also the engaged participation of all stakeholders, encompassing social, economic, and psychological aspects [9]. In this context, it is of strong practical significance to understand the game relationships and behaviors of stakeholders in recyclable resource management.

Existing studies have taken attention of stakeholder games in recyclable resource management, as to be mentioned in Sect. 3. Most of these studies focused on specific types of recyclable resources [10–18], or only included the individual phases of recycling as an integral part of recycling [19–25]. However, these insights are relatively limited, and are only applicable to explain some specific issues from certain perspectives, lacking a holistic and comprehensive understanding [26–29]. Particularly for game models, these limitations result in the resultant game equilibrium state becoming a partially stable state that does not permit a genuine stakeholder agreement. The breakdown of game models can readily result in ineffective government regulation, excessive resource consumption, heightened conflict, and other undesirable consequence. The deviation of modeling from reality has certainly posed challenges in implementing recycling.

To this end, this study systematically categorizes the problems of current stakeholder game models in recyclable resource management, and proposed a novel multi-phase stakeholder game framework, which considers all levels of stakeholders in the life cycle of recyclable resource management with thoroughness and dynamism simultaneously. The cross-sectional and longitudinal analyses of the game relationships among and within stakeholder groups were carried out to further explicate the proposed framework. By means of a systematic and understanding comprehension of stakeholder game relationships and behaviors in recyclable resource management, the research results can better steer the overall and various levels of recyclable resource management practices.

2 Methodology

In order to enhance our comprehension of the social research of the recyclable resource management system, a qualitative and critical appraisal of the published literature in associated fields were conducted, thereby creating an informative bibliography for document visualization.

2.1 Document Selection

A keyword search was performed at which called the Institute for Scientific Information (ISI) Web of Science, one of the most powerful, up-to-date, comprehensive, and widely used search engines for the analysis of interdisciplinary, peer-reviewed literature [30]. The data used for bibliometric analysis was collected by retrieving the articles which contain the keywords of Recyclable resource/Recycling/Resource and Stakeholder/Decision maker/Game theory in WOS (Web of Science). According to the title and the abstract, retrieved documents were reviewed to evaluate their suitability for inclusion in the final categorizations. However, if insufficient information was available in titles and abstracts, a full-text review was also conducted. Only articles explicitly discussing the social dimensions of recyclable resource management were selected.

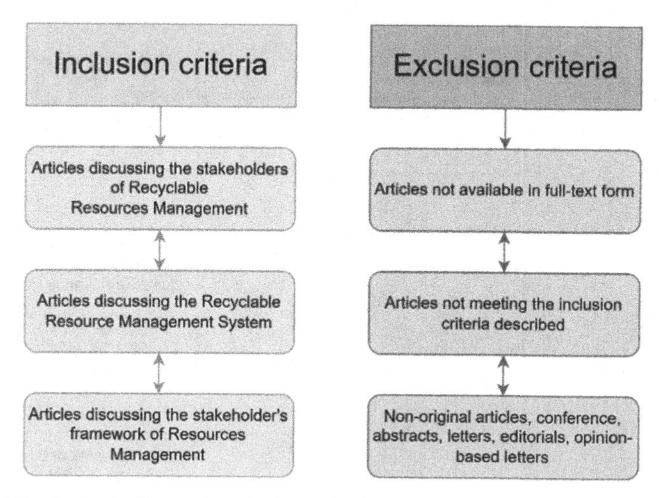

Fig. 1. Inclusion and exclusion criteria used for the article selection.

2.2 Document Preview

According to the inclusion and exclusion criteria (see Fig. 1), 569 articles were obtained and 287 articles constructing stakeholder game models were manually sorted out (as shown in Fig. 2). Most case studies discussed the behaviors of stakeholders in cooperative games regarding economic factors such as enterprise performance. However, only a few documents considered social and cultural factors such as social responsibility. We find that most of the current studies explore cooperative game and non-cooperative game models deal with scenarios separately. Instead, the whole life cycle theory shows us that multiple stages of recyclable resources are carried out simultaneously. Therefore, we strive to develop a multi-agent and multi-stage comprehensive model through systematic literature review.

3 Result

In this section, a systematical literature review approach was used to summarize the problems of existing stakeholder game models in recyclable resource management.

3.1 Results of Articles Screening

A recording form was developed to document and characterize specific details in the social dimensions related to recyclable resource management, and examine key trends and associations. The recording form begins with categories related to the general characteristics of the articles in terms of their authorship, article title, the year of publication, document type, first author affiliation, and study area.

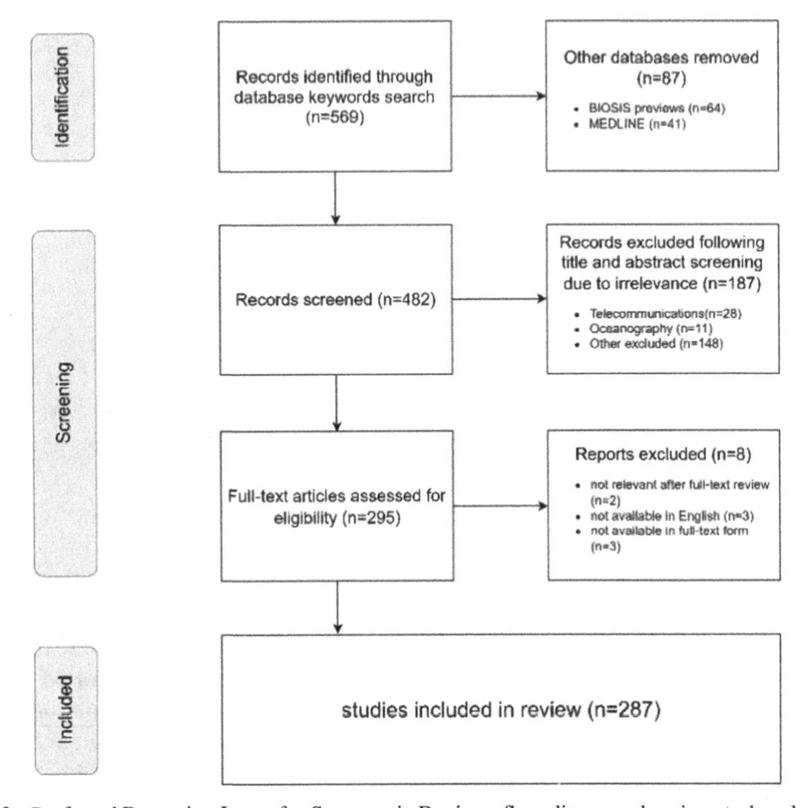

Fig. 2. Preferred Reporting Items for Systematic Reviews flow diagram showing study selection.

3.2 Problems of Existing Stakeholder Game Models in Recyclable Resource Management

The tools of Citespace and VOSviewer software were used to count the frequency of word occurrences and construct a word co-occurrence network[31–33]. As shown in Fig. 3, the word co-occurrence network constructed by VOSviewer. Overall, the words, i.e., management, sustainability, circular economy, waste management, performance, and stakeholder appear most frequently in the research themes.

The existing stakeholder game models mainly focused on various types of waste. For example, Wang et al. and Liu et al. constructed government-recycler-consumer evolutionary game model and government-recycler stochastic evolutionary game models, respectively, to explore the recycling of e-waste [13, 15]. Yang et al. analyzed the game problem of multi-agent express packaging waste recycling system [16]. Soltani used game theory to help stakeholders reach effective municipal solid waste treatment strategies [18]. However, these studies overemphasize the specificity of a certain recyclable resource type, while ignore the commonality of institutional arrangements, organizational arrangements, and stakeholder relationships in recyclable resource management. Establishing a centralized and unified management system for multiple recyclable

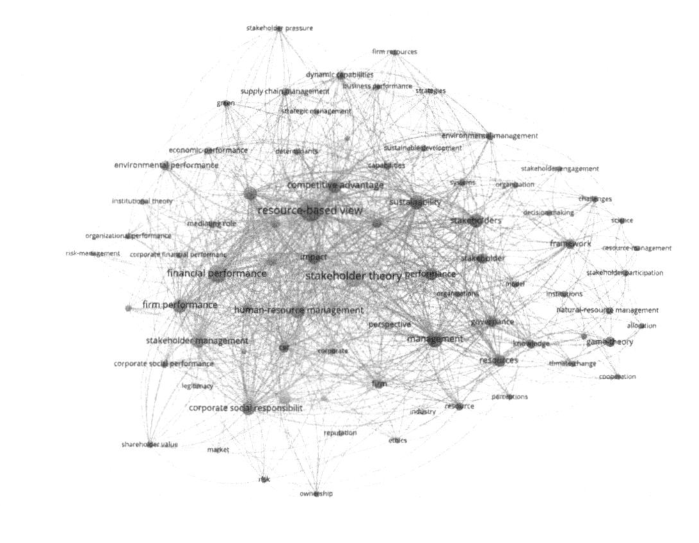

Fig. 3. Word co-occurrence network of existing stakeholder game research in recyclable resource management.

resources is conducive to the formation of scale effects and the saving of management resources, which exposes the shortcomings of the existing studies.

Most models simply treated the individual phases of recycling as an integral part of recycling. For example, Sabbaghia et al. exaggerated the stakeholder interest in the remanufacturing to that of recycling [23]. Similarly, Wang et al. and Liu et al. mainly considered the collection scenario, but summarized it with the term recycling [13, 17]. In these investigations, the resultant game equilibrium state frequently represents a partially stable state throughout the entire recycling process, which does not necessarily indicate that stakeholders can attain a consensus from the commencement to the culmination of recycling. This leads to a divergence of the modeling from the real scenario of recycling. To sum up, current game models scrutinized the stakeholder game relationships and behaviors from diverse angles for different categories of recyclable resources.

Regrettably, these insights are relatively limited and only applicable to explain some specific issues from certain perspectives. Existing models are not yet able to capture the full picture of stakeholder game relationships and behaviors in the system of recyclable resource management, which hinders the implementation of recyclable resource management [4]. Therefore, a general model is needed.

3.3 A Novel Stakeholder Game Framework for Recyclable Resource Management System

This section proposes a novel stakeholder game framework with full consideration of all levels of stakeholders in the life cycle of recyclable resource management.

As shown in Fig. 3, the proposed game framework consists of cross terms with the life cycle of recyclable resources as the horizontal axis and the stakeholder groups as the vertical axis. Integrating the views of existing literature of stakeholder game perspective and the life cycles of recycling resources [4, 16, 17, 34], the life cycle of recyclable resources should be divided into four stages which are generation, collection, processing and reuse phase (Fig. 35).

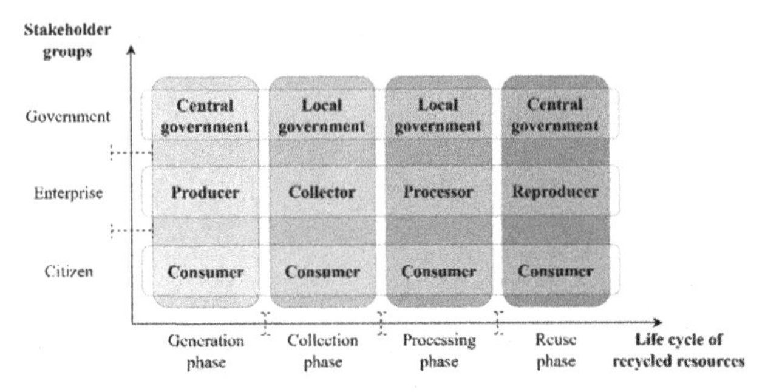

Fig. 4. Stakeholder Game Framework for Recyclable Resource Management System

The mainly aim of the generation phase is to reduce waste generation throughout the product design, production, distribution, and consumption processes. The generation phase primarily involves the federal government that establishes pertinent regulations and policies, manufacturers that create products, and consumers that purchase these products [35].

The mainly aim of the collection phase is to collect recyclable resources from consumers effectively. The collection phase mainly involves local governments that provide subsidies, collectors that gather recyclable resources, and consumers that possess recyclable resources.

The mainly aim of the processing phase is to decompose waste and extract reusable components in the right way. The processing phase primarily involves local governments that provide subsidies, processors that handle recyclable resources, and consumers that oversee processing activities.

The mainly aim of the reuse phase is to encourage manufacturers to use recyclable resources and to encourage consumers to purchase them. The reuse phase mainly involves the federal government that establishes pertinent regulations and policies, reproducers that manufacture the reproduced products, and consumers that use these reproduced products.

In summary, all the aforementioned stakeholders can be categorized into three stakeholder groups: government, businesses, and citizens [16]. To comprehensively analyze this game framework, we perform cross-sectional and longitudinal analyses of the game relationships among and within stakeholder groups, respectively.

3.4 Cross-Sectional Analysis Among Stakeholder Groups

This section analyzes the four game models among stakeholder groups corresponding to each phase.

In the Generation phase, the three parties involved in the game are the Central government, Producers, and Consumers. The central government has two strategies to

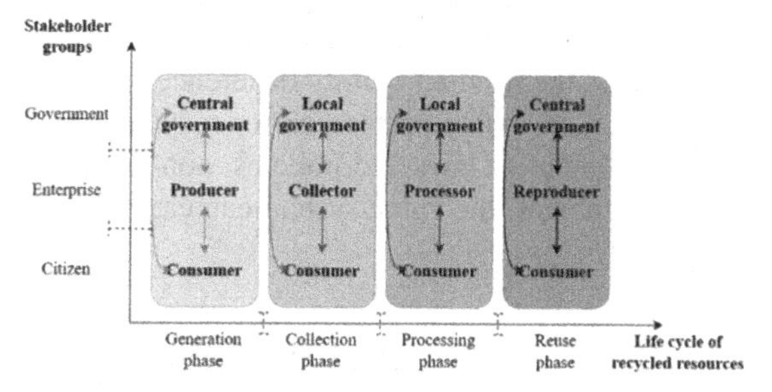

Fig. 5. Cross-sectional analysis among stakeholder groups in recyclable resource management

choose from (strict regulation and negative regulation). Producers have two strategies to choose from (waste reduction and waste non-reduction) [24], while consumers have two strategies to choose from (response and non-response). The primary factors that impact the tripartite game are the cost of government regulation, the technical and implementation costs that associated with waste reduction for producers, product prices, and consumer awareness regarding waste generation. By implementing laws and policies, reducing waste production costs, and increasing consumer awareness, the game system can quickly reach a consensus on waste reduction (Fig. 5).

In the Collection phase, Local governments, Collectors and Consumers are the three parties of the game. The strategic space for Local governments is (subsidy and non-subsidy), for Collectors is (legal collection and illegal collection), and for Consumers is (response and non-response). The main factors affecting the tripartite game are the subsidies given by Local governments to legal Collectors and Consumers, the profits of legal and illegal collection by Collectors, the fines for illegal collection by Collectors, and the proceeds from the sale of Consumers to legal and illegal collectors, respectively. When Local governments provide sufficient subsidies and fines for illegal collection are large enough, the game system converges to consensus for legal collection.

In the Processing phase, Local governments, Processors and Consumers are the three parties in the game. The strategic space for Local governments is (ex-ante prevention, ex-post governance), for Processors is (compliance processing, non-compliance processing), and for Consumers is (response, non-response). The main factors affecting the tripartite game are: the cost of environmental governance by Local governments, the technical cost and implementation cost of compliance processing by Processors, the cost of monitoring and reporting by Consumers, and the health cost paid by consumers when contaminated. To reduce the environmental governance cost of Local governments, Local governments prefer to prevent pollution beforehand by forcing disposers to comply with waste processing through policy and economic means. At the same time, Consumers monitor the behaviors of Local governments and Processors in pursuit of physical health.

In the Reuse phase, the Central government, Reproducers and Consumers are the three parties in the game. The strategic space for the Central government is (strict regulation and negative regulation), for Reproducers is (reuse of recyclable resources and utilization of first-use resources), and for Consumers is (response and non-response) [19, 23, 22]. The main factors affecting the tripartite game are: the cost of regulation by the

Central government, the cost of adopting recyclable and first-use resources by Reproducers, the price of products, the product quality difference between using recyclable and first-use resources, and Consumers preference for recyclable products [16]. With the introduction and enforcement of laws and policies, non-declining product quality, and shifting consumer preferences, the game system can easily converge to consensus for a thriving reuse industry.

3.5 Longitudinal Analysis Within Each Stakeholder Group

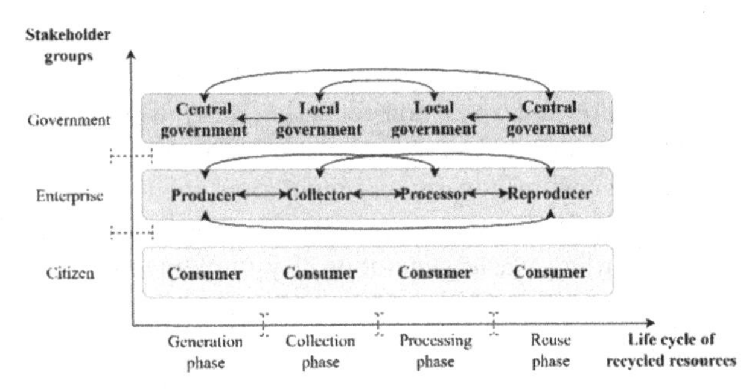

Fig. 6. Longitudinal analysis within each stakeholder group in recyclable resource management

This section analyzes the two cross-phase game models within each stakeholder group. Within the government, the Central government and Local governments are the two parties in the game [16]. The strategy space for the Central government is (strict regulation and negative regulation), and for Local governments is (active governance and passive governance). The main factors affecting the tripartite game are the cost of regulation by the Central government, the cost of governance by Local governments, the dedicated funds from the Central government to Local governments, the reward for effective governance by Local governments, and the penalty for inadequate governance by Local governments. With strong dedicated financial support and right policies on incentives and penalties, the game system will converge to consensus for proper governance (Fig. 6).

In the recycling business chain, there are four parties involved in the game Producers, Collectors, Processors, and Reproducers. The strategic options for Producers are (accountable and unaccountable), for Collectors are (accountable and unaccountable), for Processors are (accountable and unaccountable), and for Reproducers are (reusing recyclable resources and using first-use resources). The primary factors that impact the four-sided game are: the technical and implementation costs of waste reduction for Producers, the profits from legal and illegal collection for Collectors, the technical and implementation costs of compliance processing for Processors, the cost of adopting recyclable and first-use resources for Reproducers, and the willingness of businesses to cooperate. When the cost for each enterprise is significantly lower than the benefits that can be obtained from cooperation, businesses will spontaneously promote responsible recycling.

4 Conclusion

This research is the initial attempt to explain the various and varied stakeholder game connections and actions in the management of recyclable resources, use inductive methodology on the basis of which a new multi-phase stakeholder game structure for the management of recyclable resources has been suggested. In comparison to current stakeholder game research, the suggested game structure could evaluate the stakeholder games in the management of recyclable resources in a more comprehensive and complete manner, in order to more effectively lead the management of recyclable resources as a whole and at all levels. Considering the complexity of multi-stakeholder game, this paper just retains the macro content of the game model and provides a macro model framework for the situational analysis of specific cases.

There are some areas that could be improved. (1) Within the citizens, further discourse is required on how consumer interactions can encourage resource recycling. In subsequent research, techniques like intricate network analysis can be utilized to tackle this matter. (2) This study merely presents fundamental modeling ideas in the cross-sectional and longitudinal analyses owing to the article's length constraints. Upcoming research can enhance these ideas and authenticate the game outcomes by implementing modeling and simulation.

Acknowledgments. Thanks for the unity and cooperation of our team.

Disclosure of Interests. The authors have no competing interests to declare that are relevant to the content of this article.

References

1. Barnes, J.H.: Recycling: a problem in reverse logistics. J. Macromark. **2**(2), 31–37 (2016). https://doi.org/10.1177/027614678200200204
2. Silva, W.D.O., Morais, D.C.: Impacts and insights of circular business models' outsourcing decisions on textile and fashion waste management: a multi-criteria decision model for sorting circular strategies. J. Clean. Prod. **370**, 133551 (2022). https://doi.org/10.1016/j.jclepro.2022.133551
3. Ma, L., MZhang, L.: Evolutionary game analysis of construction waste recycling management in China. Resourc. Conserv. Recycl. **161**, 104863 (2020). https://doi.org/10.1016/j.resconrec.2020.104863
4. Guo, S., Chen, L.: Why is China struggling with waste classification? A stakeholder theory perspective. Resourc. Conserv. Recycl. **183**, 106312 (2022)
5. Yin, Z., Ma, J., Liu, Y., He, J., Guo, Z.: New pathway exploring the effectiveness of waste recycling policy: a quasi-experiment on the effects of perceived policy effectiveness. J. Clean. Prod. **363**, 132569 (2022). https://doi.org/10.1016/j.jclepro.2022.132569
6. Herat, S.: E-waste management in Asia pacific region: review of issues, challenges and solutions. Nat. Environ. Pollut. Technol. **20**(1), 45–53 (2021). https://doi.org/10.46488/NEPT.2021.v20i01.005
7. Hao, X., Liu, G., Zhang, X., Dong, L.: The coevolution mechanism of stakeholder strategies in the recycled resources industry innovation ecosystem: the view of evolutionary game theory. Technol. Forecast. Soc. Change. **179**, 121627 (2022)

8. Hipel, K.W., Fang, L., Wang, L.: Fair water resources allocation with application to the south Saskatchewan river basin. Can. Water Resourc. J. **38**(1), 47–60 (2013). https://doi.org/10.1080/07011784.2013.773767
9. Ma, J., Hipel, K.W.: Exploring social dimensions of municipal solid waste management around the globe - a systematic literature review. Waste Manage. **56**, 3–12 (2016)
10. Ding, L., Guo, Z., Xue, Y.: Dump or recycle? Consumer's environmental awareness and express package disposal based on an evolutionary game model. Environ. Dev. Sustain. **25**(7), 6963–6986 (2022). https://doi.org/10.1007/s10668-022-02343-1
11. Dong, Y., Li, J., Liu, T., Fan, M., Yu, S., Zhu, Y.: Evolutionary game analysis for protecting suppliers' privacy between government and waste mobile phone recycling companies: insights from prospect theory. J. Intell. Fuzzy Syst. **43**(3), 3115–3132 (2022)
12. Li, S., Sun, Q.: Evolutionary game analysis of WEEE recycling tripartite stakeholders under variable subsidies and processing fees. Environ. Sci. Pollut. Res. **30**(5), 11584–11599 (2022)
13. Liu, X., Lin, K., Wang, L.: Stochastic evolutionary game analysis of e-waste recycling in environmental regulation from the perspective of dual governance system. J. Clean. Prod. **319**, 128685 (2021)
14. Shao, Y., Deng, X., Qing, Q., Wang, Y.: Optimal battery recycling strategy for electric vehicle under government subsidy in China. Sustainability **10**(12), 4855 (2018). https://doi.org/10.3390/su10124855
15. Soltani, A., Sadiq, R., Hewage, K.: Selecting sustainable waste-to-energy technologies for municipal solid waste treatment: a game theory approach for group decision-making. J. Clean. Prod. **113**, 388–399 (2016). https://doi.org/10.1016/j.jclepro.2015.12.041
16. Su, Y., Si, H., Chen, J., Wu, G.: Promoting the sustainable development of the recycling market of construction and demolition waste: a stakeholder game perspective. J. Clean. Prod. **277**, 122281 (2020)
17. Wang, Z., Wang, Q., Chen, B., Wang, Y.: Evolutionary game analysis on behavioral strategies of multiple stakeholders in E-waste recycling industry. Resourc. Conserv. Recycl. **155**, 104618 (2020)
18. Yang, J., Long, R., Chen, H., Sun, Q.: A comparative analysis of express packaging waste recycling models based on the differential game theory. Resourc. Conserv. Recycl. **168**, 105449 (2021)
19. Askar, S.S., Al-khedhairi, A.: A remanufacturing duopoly game based on a piecewise nonlinear map: analysis and investigations. Int. J. Nonlinear Sci. Numer. Simul. **21**(6), 549–561 (2020). https://doi.org/10.1515/ijnsns-2019-0150
20. Gupta, V.K., Kaushal, R.K., Shukla, S.P.: Multi-stakeholder policy modeling for collection and recycling of spent portable battery waste. Waste Manage. Res. J. Sustain. Circ. Econ. **36**(7), 577–593 (2018)
21. Ji, Y.J., Jiao, R.J., Chen, L., Wu, C.L., Li, H.: A game theoretic model for analysis of material reuse modularity. In: 2012 IEEE International Conference on Industrial Engineering and Engineering Management (2012)
22. Li, J., Du, W., Yang, F., Hua, G.: Evolutionary game analysis of remanufacturing closed-loop supply chain with asymmetric information. Sustainability. **6**(9), 6312–6324 (2014)
23. Sabbaghi, M., Behdad, S., Zhuang, J.: Managing consumer behavior toward on-time return of the waste electrical and electronic equipment: a game theoretic approach. Int. J. Prod. Econ. **182**, 545–563 (2016). https://doi.org/10.1016/j.ijpe.2016.10.009
24. Zhao, X., Bai, X., Fan, Z., Liu, T.: Game analysis and coordination of a closed-loop supply chain: perspective of components reuse strategy. Sustainability **12**(22), 9681 (2020). https://doi.org/10.3390/su12229681
25. Zhou, Z.F.: Research on pricing decision of multi-level remanufacturing reverse supply chain based on Stackelberg game. Appl. Mech. Mater. **220–223**, 290–293 (2012). https://doi.org/10.4028/www.scientific.net/AMM.220-223.290

26. Bekius, F., Meijer, S., Thomassen, H.: A real case application of game theoretical concepts in a complex decision-making process: case study ERTMS. Group Decis. Negot. **31**(1), 153–185 (2021). https://doi.org/10.1007/s10726-021-09762-x
27. Nagayama, D., Horita, M.: A network game analysis of strategic interactions in the international trade of Russian natural gas through Ukraine and Belarus. Energy Econ. **43**, 89–101 (2014)
28. Ma, J., Hipel, K.W., Hanson, M.L., Cai, X., Liu, Y.: An analysis of influencing factors on municipal solid waste source-separated collection behavior in Guilin, China by Using the Theory of Planned Behavior. Sustain. Cities Soc. **37**, 336–343 (2018)
29. Ma, J., Hipel, K.W., Hanson, M.L.: Public participation in municipal solid waste source-separated collection in Guilin, China: status and influencing factors. J. Environ. Planning Manage. **60**(12), 2174–2191 (2017)
30. Jacso, P.: As we may search - Comparison of major features of the Web of Science, Scopus, and Google Scholar citation-based and citation-enhanced databases. Curr. Sci. **89**(9), 1537–1547 (2005)
31. Chen, C.: CiteSpace II: detecting and visualizing emerging trends and transient patterns in scientific literature. J. Am. Soc. Inform. Sci. Technol. **57**(3), 359–377 (2006)
32. Sousa, M.J., Rocha, Á.: Decision-making and negotiation in innovation; research in information science. Group Decis. Negot. **30**(2), 267–275 (2020)
33. van Eck, N.J., Waltman, L.: Software survey: VOS viewer, a computer program for bibliometric mapping. Scientometrics **84**(2), 523–538 (2009). https://doi.org/10.1007/s11192-009-0146-3
34. Gu, Y., et al.: Environmental performance analysis on resource multiple-life-cycle recycling system: Evidence from waste pet bottles in China. Resourc. Conserv. Recycl. **158**, 104821 (2020)
35. Nie, Z., Gao, F., Gong, X., Wang, Z., Zuo, T.: Recent progress and application of materials life cycle assessment in China. Progr. Nat. Sci. Mater. Int. **21**(1), 1–11 (2011). https://doi.org/10.1016/S1002-0071(12)60016-1
36. Chu, X., Shi, Z., Yang, L., Guo, S.: Evolutionary game analysis on improving collaboration in sustainable urban regeneration: a multiple-stakeholder perspective. J. Urban Planning Dev. **146**(4), 04020046 (2020). https://doi.org/10.1061/(ASCE)UP.1943-5444.0000630

Hybrid Evolutionary Approach to Team Building using PROMETHEE II

Georgios Stavrou[1] (✉), Panagiotis Adamidis[2], Jason Papathanasiou[3], and Konstantinos Tarabanis[3]

[1] Department of Information and Electronic Engineering, University of Macedonia, 57400 Thessaloniki, Greece
gstavr@uom.edu.gr
[2] International Hellenic University, Egnatias 156, 54636 Thessaloniki, Greece
adamidis@ihu.gr
[3] University of Macedonia, Egnatias 156, 54636 Thessaloniki, Greece
{jasonp,kat}@uom.edu.gr

Abstract. Multi-Criteria Decision Making (MCDM) methods, such as PROMETHEE II, play a crucial role in complex decision-making scenarios, including team formation. However, they face scalability challenges as the number of criteria and options increases. This paper introduces a novel Hybrid Evolutionary Algorithm integrated with PROMETHEE II, specifically designed for team formation. This hybrid approach combines the exploration power of evolutionary algorithms and the decision-making capabilities of PROMETHEE II, aiming to improve both performance and scalability in decision-making processes. Initial experiments demonstrate significant improvements in both solution quality and scalability compared to existing methods facing similar challenges. This research enables the creation of more efficient and effective team formation in complex decision-making scenarios.

Keywords: Multi-Criteria Decision Making · Team Formation · Evolutionary Algorithms · PROMETHEE II · Algorithmic Efficiency · Computational Optimization · Team Composition Analysis · Decision Support Systems

1 Introduction

In the complex landscape of decision-making, fueled by the exponential growth of data and diverse variables, Multi-Criteria Decision Making (MCDM) methods stand as critical tools for navigating these challenges. Among the array of MCDM methods, PROMETHEE II is distinguished by its adaptability and robustness, making it particularly suited for nuanced evaluations such as team formation. Yet, traditional applications of PROMETHEE II, and similar MCDM approaches, grapple with limitations in scalability and computational efficiency when confronted with large datasets and an extensive array of decision-making criteria [1–8].

M. Campos Ferreira et al. (Eds.): GDN 2024, LNBIP 509, pp. 38–48, 2024.
https://doi.org/10.1007/978-3-031-59373-4_4

Addressing these challenges, this study introduces a novel Hybrid Evolutionary Algorithm that synergizes with PROMETHEE II, explicitly designed to enhance the dynamics of team formation. This innovative approach meticulously balances considerations of expertise, personality compatibility, and specific project requirements—dimensions where conventional methods may falter, especially under the pressures of scalability and performance [9–13].

Capitalizing on the strengths of Evolutionary Algorithms (EAs) for their unparalleled efficiency in navigating expansive solution spaces, this research endeavors to surmount the scalability and efficiency hurdles observed in traditional decision-making frameworks [4, 14, 15]. While the fusion of EAs with MCDM techniques is not a new concept, prior implementations have not fully exploited the synergistic potential inherent in such combinations, particularly within the context of large-scale team formation endeavors. Our methodology distinguishes itself through the strategic application of EAs to augment the decision-making prowess of PROMETHEE II, directly confronting scalability and performance challenges [4, 16].

This paper is organized to underscore both the theoretical and practical merits of merging EAs with MCDM methods, with a special focus on PROMETHEE II. It delves into the intricacies of our Hybrid Evolutionary Algorithm integrated with PROMETHEE II, elucidating its operational framework and delineating its advantages over conventional approaches. Concentrating on the domain of team formation, a scenario emblematic of the need to harmonize diverse and at times, competing criteria, this investigation demonstrates the effectiveness and applicability of our proposed strategy in simulated environments that closely replicate the complexities of real-world applications.

2 Background and Related Work

The realm of decision-making in complex environments has been significantly advanced by the development of Multi-Criteria Decision Making (MCDM) methods. Among these methods, PROMETHEE II is notable for its versatility in handling a broad spectrum of decision-making scenarios, from corporate strategy to environmental management [7]. Its adaptability across various fields demonstrates its capability to address diverse decision matrices effectively[17]. Despite its strengths, the scalability and computational efficiency of conventional PROMETHEE II implementations pose challenges when applied to large datasets or when evaluating a vast number of criteria.

Recent advancements in Evolutionary Algorithms (EAs) have shown potential in addressing these scalability concerns. EAs, renowned for their proficiency in navigating complex solution spaces, offer a promising avenue for enhancing MCDM methods [18, 19]. Their adaptability and evolutionary nature make them particularly suited for scenarios where traditional algorithms falter due to the complexity or size of the problem space [20].

The synergy between EAs and MCDM, while recognized, has been underexplored in the context of team formation. This research aims to bridge this gap by integrating EAs with PROMETHEE II, focusing on team formation's dynamic and multifaceted nature. Team formation involves more than just aggregating individual skills; it requires a nuanced consideration of expertise, personality compatibility, and project-specific demands. Traditional MCDM approaches often struggle to capture these complexities, especially in larger, more diverse settings.

This paper presents a novel approach by combining the exploratory strength of EAs with the decision-making precision of PROMETHEE II, aiming to enhance both scalability and decision quality in team formation. By leveraging this hybrid model, we explore how synergistic effects between team members can be quantified and optimized, addressing a critical gap in existing literature [6, 12, 21–23].

Our contribution is twofold: firstly, to the academic discourse on MCDM and EA integration, offering a novel methodology for complex decision-making scenarios. Secondly, to practical applications in team formation, demonstrating how this hybrid approach can lead to more effective and efficient team assembly processes. This research not only explores the theoretical underpinnings of such an integration but also presents simulated scenarios that reflect real-world complexities, underscoring the potential of this method in both academic and practical domains [4, 24, 25].

3 Methodology

To effectively tackle scalability and efficiency challenges in team formation, our methodology combines a Hybrid Evolutionary Algorithm with PROMETHEE II. This integration leverages the evolutionary algorithm's capability for broad solution space exploration alongside PROMETHEE II's precise decision-making framework. This synergy enables a more dynamic assessment of potential team configurations, surpassing traditional MCDM methods in handling complex scenarios with diverse criteria.

Data and Simulation Environment

To evaluate performance across diverse project scenarios, our study developed a dataset of 3,000 simulated profiles, each reflecting the diversity of skills and personalities based on the Myers-Briggs Type Indicator (MBTI). This diverse dataset ensures a comprehensive test of the algorithm's adaptability and performance, considering the complexity of both team dynamics and project requirements [28–29]. It encompasses a range of skills (more than 150 skills) and MBTI personality traits and matching scores, reflecting real-world professional diversity. This allows the exploration of how these variables influence team effectiveness. The simulated projects vary in size and complexity, with specific requirements for skills and personality compatibility, allowing us to assess the algorithm's effectiveness in forming teams under diverse conditions. For example, each virtual project specifies a set of desired skills, with associated minimum and maximum skill values. This range ensures that selected team members possess skills within the defined spectrum, aligning with project needs.

Experiment Procedure

Our experimental framework simulates real-world constraints by incorporating a 300-s time limit, a stagnation check at 200 generations without improvement, and a 1200-generation cap. These parameters are set to mimic real-world project urgency and computational limitations, ensuring the algorithm's performance is evaluated under conditions that reflect practical scenarios of team formation. This approach balances the need for comprehensive exploration against the imperative for timely decision-making.

Within this context, the evolutionary algorithm was tailored with variable settings to address distinct project requirements, emphasizing team size, skill set diversity and personality compatibility scores. We engaged with two project scenarios: the first demanded a 5-member team equipped with 4 specific skill sets and a personality weight of 10, while the second required a 4-member team with 6 skill sets and a personality weight of 20. Skill level requisites for these projects ranged from 30 to 86 and 50 to 70, respectively, showcasing varied project needs. To explore the algorithm's adaptability, population sizes of 30 and 200 were tested, reflecting different operational scales. Furthermore, each algorithm setting was executed more than 20 times for each project, ensuring robustness and reliability in the results obtained.

Our experimental design also included 12 evolutionary settings adjustments, particularly focusing on mutation rates between 0.015 to 0.02 and variations in population size. This ensured every chromosome underwent crossover, facilitating the generation of new offspring for subsequent evaluation. The fitness function, pivotal to our experiment, was devised to measure potential team cohesion, alignment with project-specific skill requirements, and personality compatibility. This multi-dimensional evaluation, leveraging both skill sets and personality weights, underscores the algorithm's sophistication in optimizing team configurations under stringent time constraints and the dynamism of project demands.

Algorithmic Framework for PrometheeEA

Our hybrid algorithm not only evaluates potential team solutions against virtual benchmarks for skill requirements but also assesses the synergistic effects of individuals within teams, going beyond simple skill compatibility and dynamics. This assessment is facilitated by integrating a nuanced measure of team synergy into the fitness function. Specifically, the algorithm quantifies synergy through a combination of metrics that capture the enhanced productivity and creativity resulting from effective interpersonal interactions and complementary skill sets. These metrics include collaborative efficiency, where team members' ability to work together seamlessly is evaluated, and the diversity of thought, which assesses how varied perspectives contribute to innovative problem-solving.

Through comparative analysis with virtual teams, the algorithm leverages the analytical capabilities of PROMETHEE II to identify team configurations that not only meet skill requirements but also optimize team synergy. The insights gained from evaluating these synergistic effects are pivotal in refining the solution pool, guiding the evolutionary algorithm to focus on the most promising candidates that exhibit both high skill compatibility and exceptional team synergy.

Incorporating synergy into the decision-making process ensures that the selected teams are not just qualified on paper but are also likely to perform exceptionally well in real-world scenarios. This integrated approach, combining the exploratory power of evolutionary algorithms with the evaluative precision of PROMETHEE II, represents a novel solution to the complex challenge of team composition, aiming to significantly improve the efficiency and effectiveness of forming teams for complex projects.

Algorithmic Framework for Promethee II

The integration of the PROMETHEE II within the evolutionary framework serves two key purposes: maintaining consistency in population size and facilitating recombination to effectively explore a broader solution space guided by PROMETHEE II evaluations. The fitness function, a key component of this process, specifically applies PROMETHEE II across the entire population, utilizing the project's criteria and weights to assess each solution's viability. Directly applying PROMETHEE II to the vast solution space of all possible team configurations would be computationally infeasible. Thus, the evolutionary algorithm becomes instrumental in condensing this space into a manageable subset for subsequent PROMETHEE II analysis.

Through recombination and mutation processes, the evolutionary algorithm not only narrows down the initial solution space but also iteratively refines and expands it. This ensures that at any given iteration, PROMETHEE II is applied to a specific and manageable set of solutions, making the exploration process both targeted and comprehensive. Over time, this iterative refinement allows for the exploration of a diverse range of potential team configurations, enhancing the algorithm's ability to identify optimal team compositions.

This iterative exploration is vital for the effective evaluation of team compositions, as it dynamically adjusts the solution space to fit within the project's constraints. Moreover, the preference functions employed in the PROMETHEE II analysis are specifically designed to favor higher values, ensuring that if one alternative scores higher than another on any given criterion, it is preferred, reflecting the beneficial nature of that criterion. This approach guarantees that the evaluation process is closely aligned with the project's goal of identifying superior team configurations, by systematically favoring solutions that meet or exceed the project requirements.

Operational Nuances

The hybrid algorithm balances the exploration of diverse solutions with the refinement toward optimal outcomes. It capitalizes on the evolutionary algorithm's ability for broad search, preventing premature convergence on suboptimal solutions. At the same time, the analytical precision of PROMETHEE II guides this exploration, ensuring that efforts are concentrated on the most promising areas of the solution space. This is achieved through the use of preference functions that evaluate and rank potential solutions based on a set of carefully selected criteria, reflecting the complexities and specific requirements of team effectiveness.

Flexibility is crucial, enabling the algorithm to tailor its parameters and evaluation criteria to unique project demands and specific team characteristics. This adaptability is evident in its ability to adjust search parameters based on project size, team diversity, and required skill sets, ensuring applicability across various team formation scenarios. Such versatility underscores the algorithm's utility in navigating the various challenges inherent in team composition.

While the hybrid algorithm is robust and adaptable, it is important to acknowledge the challenges encountered in its implementation. These include the computational demands of managing a large solution space and the intricacies involved in tuning parameters for different project contexts. Additionally, defining appropriate criteria that accurately capture the nuances of team dynamics and project requirements can be challenging, yet is crucial for the success of the algorithm.

By addressing these operational nuances, the hybrid algorithm demonstrates its capacity to effectively balance exploration and optimization, adapt to diverse project needs, and navigate the complexities of team formation, making it a valuable tool for achieving optimal team configurations.

4 Results

The comparative evaluation of PrometheeEA and PrometheeII algorithms in the context of team formation focused on four key metrics: BestFitness Value, Runtime, Generation Count, and HasRequiredSkillSet (ensuring teams possess all required skills). The analysis, accompanied by informative graphs, examines the performance of each algorithm.

Best Fitness Value: The first graph reveals a subtle but consistent trend: PrometheeII slightly outperforms PrometheeEA in terms of the average fitness value of teams formed. This trend suggests that PrometheeII may be more adept at optimizing the team composition in alignment with the predefined fitness criteria. This might be attributed to PrometheeII's internal mechanisms being more finely tuned to team synergy and project requirements [Fig. 1]. It's important to note that PrometheeEA prematurely terminated not due to a lack of potential improvement, but rather a preset generation limit. This suggests that PrometheeEA could likely achieve better results with more computational time. In future studies, we intend to explore the performance of PrometheeEA over extended generations.

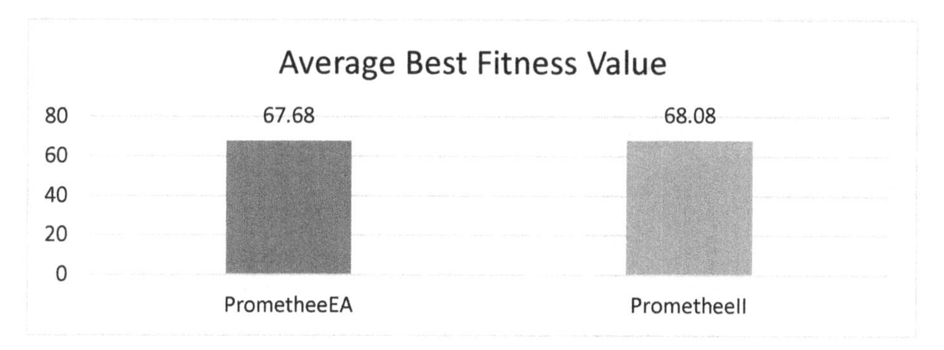

Fig. 1. Average Best Fitness Value

Runtime: The second graph highlights a significant difference in runtime: PrometheeII takes considerably longer on average, suggesting a more complex and potentially thorough evaluation process. While beneficial for team optimization, this raises concerns about its suitability in time-sensitive scenarios. In contrast, PrometheeEA's faster runtime indicates a more streamlined approach, making it more suitable for situations where rapid team formation is crucial [Fig. 2].

Generation Count: The third graph explores the number of generations, reflecting the distinct search strategies of each algorithm. To avoid random searches,

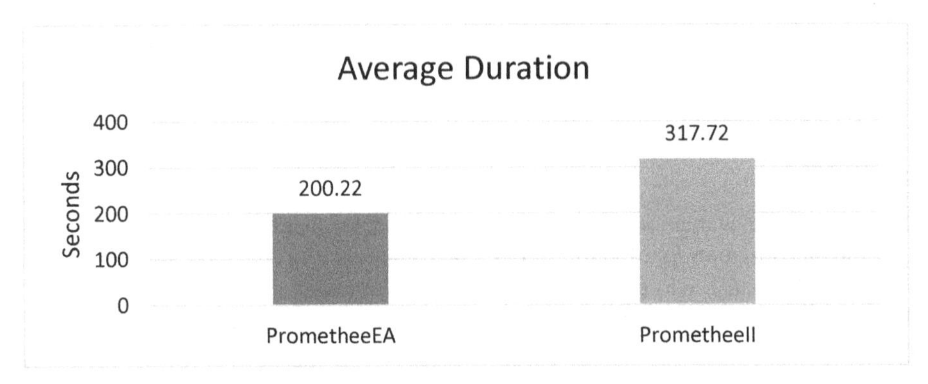

Fig. 2. Average runtime (seconds)

PrometheeII leverages evolutionary algorithms primarily for focused exploration, guided by PrometheeII's evaluation for efficient convergence.

PrometheeEA's higher generation count reflects a broader search of the solution space. This approach, though broader, is designed to discover diverse and potentially more effective team configurations. Despite its comprehensiveness, PrometheeEA maintains efficiency and the potential to identify unconventional yet highly optimized teams, contrasting with PrometheeII's more targeted approach that might prioritize convergence over broader exploration [Fig. 3].

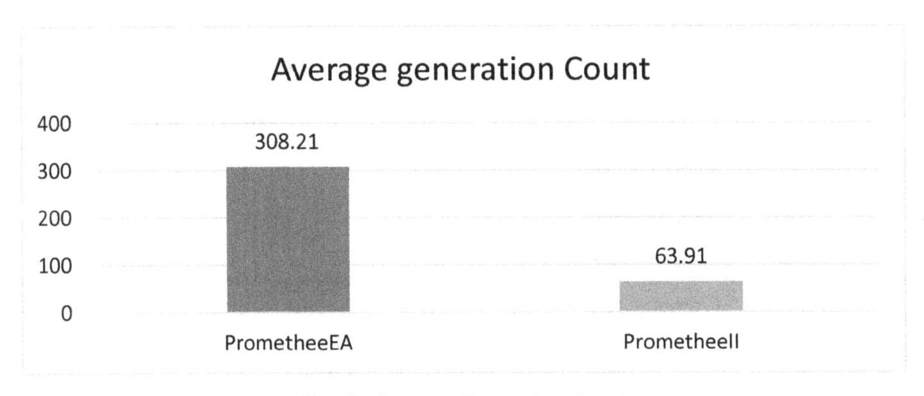

Fig. 3. Average Generation Count

Required Skill Set: The final graph examines the success rate of forming teams with the required skill sets, a crucial metric in real-world scenarios where meeting these requirements is essential. PrometheeII demonstrates a slight edge in this aspect, potentially due to its focus on optimized team formation [Fig. 4].

Enhanced Solution Space Exploration: The new algorithm excels at exploring the solution space, covering up to 3 times more ground than PrometheeII. This makes it particularly adept at identifying diverse and effective team configurations in complex projects [Fig. 5 left].

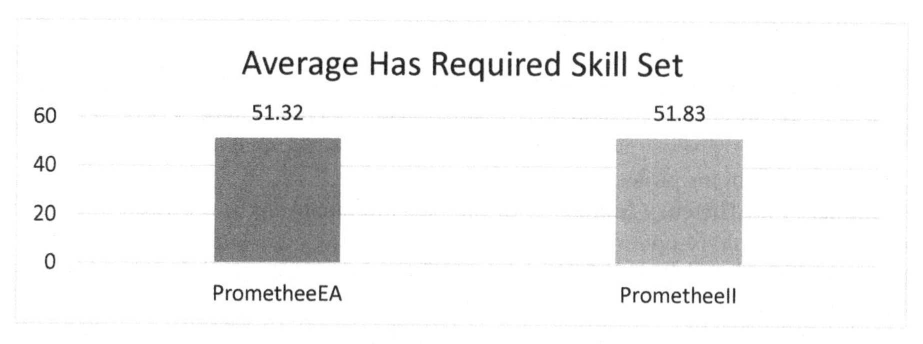

Fig. 4. Average Teams With the required Skillset for the project

Performance in Large-scale Problems: Unlike traditional MCDM methods that struggle with performance in large-scale problems, the new algorithm achieves solution quality comparable to PrometheeII while significantly improving efficiency. It reduces computation time by approximately 37%, effectively addressing scalability issues common in conventional MCDM methods [Fig. 5 right].

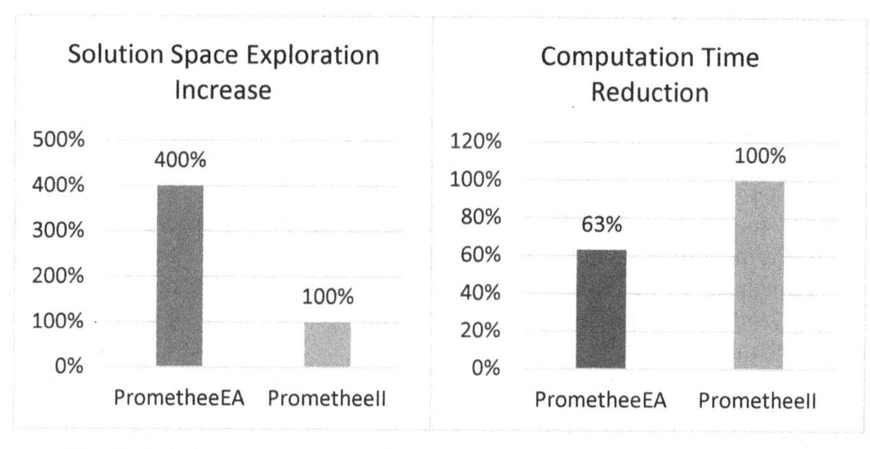

Fig. 5. Solution Space Exploration (left), Computation Time Reduction (right)

5 Conclusion

Our investigation of PrometheeEA and PrometheeII algorithms for team formation yielded valuable insights into overcoming scalability and efficiency challenges in large-scale Multi-Criteria Decision Making (MCDM). A key advancement is the development of a novel algorithm rivals PrometheeII in solution quality while significantly enhances performance efficiency, reducing computation times by approximately 37%. This breakthrough addresses the scalability limitations of traditional MCDM methods and suggests a promising avenue for future algorithmic development in complex project management.

Key Findings and Practical Implications:

- **Enhanced Solution Space Exploration**: The ability of the novel algorithm to explore the solution space 3 times more extensively than PrometheeII underscores its potential to discover diverse and optimal team configurations, making it valueable for navigating complex project requirements.
- **Quality and Efficiency Balance**: Our findings illuminate a crucial balance between the solution quality optimized by PrometheeII and the computational efficiency of the novel algorithm. This balance is essential for time-constrained projects, offering a viable alternative to traditional approaches.
- **Adaptability in Application**: The proposed algorithm demonstrates exceptional adaptability, efficiently processing a wide range of project requirements efficiently. This makes it a valuable tool for large-scale projects that demand rapid and effective team formation.

These contributions significantly enhance the decision-making processes in complex project management, proposing a methodology that carefully considers both team composition quality and algorithmic efficiency.

5.1 Limitations and Directions for Future Research

This study, while advancing the field, acknowledges limitations such as its focus on specific algorithms and the use of simulated data. Future research will extend beyond these boundaries to include a wider array of algorithms and real-world validation. There is also a keen interest in delving deeper into the complexities of team formation in real-world settings.

Planned future research includes investigating the impact of extended generation limits for PrometheeEA. While PrometheeII demonstrated a slight advantage in this study, PrometheeEA's performance was potentially limited by the preset generation limit. Analyzing PrometheeEA's performance with extended generations is crucial to determine its full potential.

Future research also includes exploring methodologies like TOPSIS and both Multi-Objective and Single Objective Evolutionary Algorithms.

We also intend to integrate real-world data and dynamic team formation. Further exploration may involve using real-world data reflecting team member attributes and project specificities. Additionally, investigating how these algorithms adapt to evolving team dynamics and project requirements would offer valuable insights for practical applications.

These efforts aim to:

- Broaden problem-solving strategies, enhancing performance across various project scenarios.
- Undertake detailed comparative analyses to identify algorithms best suited for distinct team formation challenges.
- Expand validation processes to ensure findings are robust and applicable in practical settings.

By addressing these areas, future research will develop more adaptable, efficient, and effective team formation strategies, meeting the evolving demands of real-world projects.

References

1. Taherdoost, H., Madanchian, M.: Multi-criteria decision making (MCDM) methods and concepts. Encyclopedia **3**(1), 77–87 (2023). https://doi.org/10.3390/encyclopedia3010006
2. Sahoo, S.K., Goswami, S.S.: A comprehensive review of multiple criteria decision-making (MCDM) methods: advancements, applications, and future directions. Decis. Making Adv. **1**(1), 25–48 (2023). https://doi.org/10.31181/dma1120237
3. Dhurkari, R.K.: MCDM methods: practical difficulties and future directions for improvement. RAIRO-Oper. Res. **56**(4), 2221–2233 (2022). https://doi.org/10.1051/ro/2022060
4. Yu, X., Lu, Y., Yu, X.: Evaluating multiobjective evolutionary algorithms using MCDM methods. Math. Probl. Eng. **2018**, 1–13 (2018). https://doi.org/10.1155/2018/9751783
5. Mardani, A., Jusoh, A., Nor, K.M.D., Khalifah, Z., Zakwan, N., Valipour, A.: Multiple criteria decision-making techniques and their applications – a review of the literature from 2000 to 2014. Econ. Res. Ekonomska Istraživanja **28**(1), 516–571 (2015). https://doi.org/10.1080/1331677X.2015.1075139
6. Dadelo, S., Turskis, Z., Zavadskas, E.K., Dadeliene, R.: Multi-criteria assessment and ranking system of sport team formation based on objective-measured values of criteria set. Expert Syst. Appl. **41**(14), 6106–6113 (2014). https://doi.org/10.1016/j.eswa.2014.03.036
7. Brans, J.-P., Mareschal, B.: Promethee methods. In: Greco, S., Ehrgott, M., Figueria, J. (eds.) Multiple Criteria Decision Analysis: State of the Art Surveys. ISORMS, vol. 78, pp. 163–186. Springer, New York (2005). https://doi.org/10.1007/0-387-23081-5_5
8. Dyer, J.S., Fishburn, P.C., Steuer, R.E., Wallenius, J., Zionts, S.: Multiple criteria decision making, multiattribute utility theory: the next ten years. Manage. Sci. **38**(5), 645–654 (1992). https://doi.org/10.1287/mnsc.38.5.645
9. Boix-Cots, D., Pardo-Bosch, F., Pujadas, P.: A systematic review on multi-criteria group decision-making methods based on weights: analysis and classification scheme. Inf. Fusion **96**, 16–36 (2023). https://doi.org/10.1016/j.inffus.2023.03.004
10. Hong, W.-J., Yang, P., Tang, K.: Evolutionary computation for large-scale multi-objective optimization: a decade of progresses. Int. J. Autom. Comput. **18**(2), 155–169 (2021). https://doi.org/10.1007/s11633-020-1253-0
11. Ma, J., Chang, F., Yu, X.: Large-scale evolutionary optimization approach based on decision space decomposition. Front. Energy Res. **10**, 926161 (2022). https://doi.org/10.3389/fenrg.2022.926161
12. Zhang, Y., Tian, Y., Zhang, X.: A comparison study of evolutionary algorithms on large-scale sparse multi-objective optimization problems. In: Ishibuchi, H., et al. (eds.) Evolutionary Multi-Criterion Optimization: 11th International Conference, EMO 2021, Shenzhen, China, March 28–31, 2021, Proceedings, pp. 424–437. Springer International Publishing, Cham (2021). https://doi.org/10.1007/978-3-030-72062-9_34
13. Okola, I., Omulo, E.O., Ochieng, D.O., Ouma, G.: A comparison of evolutionary algorithms on a large scale many-objective problem in food–energy–water Nexus. Results Control Optim **10**, 100195 (2023). https://doi.org/10.1016/j.rico.2022.100195
14. de Almeida, A.T., Geiger, M.J., Morais, D.C.: Challenges in multicriteria decision methods. IMA J. Manage. Math. **29**(3), 247–252 (2018). https://doi.org/10.1093/imaman/dpy005

15. Chiu, C.-C., Zhang, S., Lin, J.T., Zhen, L., Huang, E.: Improving the efficiency of evolutionary algorithms for large-scale optimization with multi-fidelity models. In: 2016 Winter Simulation Conference (WSC), pp. 815–826, September 2016. https://doi.org/10.1109/WSC.2016.7822144
16. Coello, C.A.C., Lamont, G.B., Van Veldhuizen, D.A.: Evolutionary Algorithms for Solving Multi-Objective Problems. Genetic and Evolutionary Computation Series. Springer US, Boston, MA (2007). https://doi.org/10.1007/978-0-387-36797-2
17. Behzadian, M., Otaghsara, S., Yazdani, M., Ignatius, J.: A state-of the-art survey of TOPSIS applications. Expert Syst. Appl. **39**, 13051–13069 (2012). https://doi.org/10.1016/j.eswa.2012.05.056
18. Holland, J.H.: Genetic algorithms. Sci. Am. **267**(1), 66–72 (1992). https://doi.org/10.1038/scientificamerican0792-66
19. Goldberg, L.R.: An alternative 'description of personality': The Big-Five factor structure. J. Pers. Soc. Psychol. **59**(6), 1216–1229 (1990). https://doi.org/10.1037/0022-3514.59.6.1216
20. Eiben, A.E., Smith, J.E.: Introduction to Evolutionary Computing. Natural Computing Series. Springer, Berlin, Heidelberg (2015). https://doi.org/10.1007/978-3-662-44874-8
21. Coello, C.A.C., Lamont, G.B., Veldhuizen, D.A.V.: Evolutionary Algorithms for Solving Multi-Objective Problems. Second Edn. Springer, New York (2007)
22. Gazawa, F.G., Damakoa, I.: An evolutionary algorithm coupled to an outranking method for the multicriteria shortest paths problem. Am. J. Oper. Res. **9**(3), 3 (2019). https://doi.org/10.4236/ajor.2019.93007
23. Zhang, Q., Li, H.: MOEA/D: a multi-objective evolutionary algorithm based on decomposition. IEEE Trans. Evolut. Comput. **11**, 712–731 (2008). https://doi.org/10.1109/TEVC.2007.892759
24. Cannonier, C., Smith, K.: Do crib sheets improve student performance on tests? Evidence from principles of economics. Int. Rev. Econ. Educ. **30**, 100147 (2019). https://doi.org/10.1016/j.iree.2018.08.003
25. Li, M., Kim, D.: One wiki, two groups: dynamic interactions across ESL collaborative writing tasks. J. Second. Lang. Writ. **31**, 25–42 (2016). https://doi.org/10.1016/j.jslw.2016.01.002
26. "USING MYERS-BRIGGS TYPE INDICATOR (MBTI) FOR ASSESSMENT SUCCESS OF STUDENT GROUPS IN PROJECT BASED LEARNING. In: Proceedings of the 2nd International Conference on Computer Supported Education, Valencia, Spain, pp. 156–160. SciTePress - Science and Technology Publications (2010). https://doi.org/10.5220/0002859901560160
27. Zhang, L., Zhang, X.: Multi-objective team formation optimization for new product development. Comput. Ind. Eng. **64**(3), 804–811 (2013). https://doi.org/10.1016/j.cie.2012.12.015

Preference Modeling for Group Decision and Negotiation

The Enhanced TOPSIS with Application to the Evaluation of Negotiation Offers Outside Feasible Negotiation Space

Tomasz Wachowicz[1(✉)] and Ewa Roszkowska[2]

[1] University of Economics in Katowice, 1 Maja 50, 40-287 Katowice, Poland
tomasz.wachowicz@uekat.pl

[2] Bialystok University of Technology, Wiejska 45A, 15-351 Bialystok, Poland
e.roszkowska@pb.edu.pl

Abstract. The work aims to introduce an enhanced TOPSIS (Technique for Order of Preference by Similarity to Ideal Solution) method incorporating individual aspiration and reservation levels as reference alternatives for evaluating the decision matrix. The proposed algorithm can effectively be used in multi-issue negotiation support, allowing for the evaluation of negotiation offers outside the negotiation space. It includes scenarios where resolution levels of certain issues fall above the aspiration or below the reservation level. Two approaches for comparing such offers and reservation or aspiration levels identified in deterministic negotiation space are introduced—non-compensatory and compensatory. The latter approach replaces the distance used in the original TOPSIS mechanism with two notions of positive and negative deviations that measure how much an alternative outperforms or is outperformed by aspiration and reservation levels. A numerical example is provided to illustrate the application of an enhanced TOPSIS mechanism in the pre-negotiation preparation of the negotiation offer scoring system. Its major advantage, the stability of the scoring system for new offers submitted during the actual negotiation phase, is additionally assessed through a simulation study of deficiencies of classic TOPSIS.

Keywords: Multi-issue negotiation · Negotiation scoring system · TOPSIS · Aspiration level · Reservation level · Stable scoring

1 Introduction

Evaluating negotiation offers is crucial in multi-issue negotiations, where negotiators must simultaneously compare a series of trade-offs between all negotiation issues [17, 21, 26]. To make such evaluation easier, they are suggested to build a negotiation scoring system based on their individual preferences during the pre-negotiation preparation phase. With such a scoring system, the negotiators can be effectively supported during the actual negotiation and post-negotiation phases, not only in directly evaluating offers and counteroffers. It also helps to measure the reciprocity of concessions, visualize the negotiation progress and dynamics, verify the effectiveness of negotiated compromises,

M. Campos Ferreira et al. (Eds.): GDN 2024, LNBIP 509, pp. 51–64, 2024.
https://doi.org/10.1007/978-3-031-59373-4_5

and suggest potential improvements [26]. Since the multi-issue negotiation problem analyzed by a single negotiator resembles a standard multiple criteria decision-making (MCDM) problem with a multitude of alternatives, multiple criteria decision-aiding methods, such as SAW [2], SMART [11], AHP [2, 3, 14], TOPSIS [2, 18, 24], ELECTRE TRI [23], MARS [6, 26], or UTA [10, 25], can be employed to build negotiation scoring system. Among those methods, TOPSIS seems particularly interesting since it focuses on evaluating alternatives compared to the reference points. Such a philosophy of offer evaluation is very close to the negotiation context, where the parties often define their aspiration (targets) and reservation (break-off) levels. For this reason, TOPSIS was earlier explored and recognized as a useful negotiation support technique [2, 18, 24]. However, certain drawbacks and limitations can be identified when implementing TOPSIS to negotiation support. In the classical TOPSIS approach, reference points, i.e., ideal and anti-ideal alternatives, are derived from the set of compared alternatives. They are defined as max or min options from all evaluated offers, depending on the type of the criterion. It limits the scoring capability of the scoring system built this way to the offers from a predefined set of acceptable solutions only. However, during negotiations, counterparts may present offers beyond this set due to differences in their perceptions of feasibility based on diverse preferences and goals. The fact that one party deems certain offers unacceptable does not prevent their submission by counterparts. For instance, the negotiator may assume the feasible space for negotiated price between $5 and $10 and assume TOPSIS positive and negative single-criterion ideal solutions represent these values, while during the negotiation process she may receive an offer from the counterpart suggesting a price $3. Thus, our focal negotiator should be capable of evaluating these offers to understand their true implications. Additionally, the time pressure effect [12] may lead to a successive decrease in the reservation level from round to round. Therefore, the reliable evaluation of offers that fall below the original reservation level becomes increasingly crucial. While some considerations for offers beyond the initial negotiation space were explored using TOPSIS and fuzzy numbers [18], explicit solutions are lacking when applying classic TOPSIS with crisp scalar ratings for negotiation support. The latter is the easiest and most intuitive way of defining negotiation problems by negotiators.

In this paper, we propose an extended TOPSIS algorithm by introducing external reference points that reflect aspiration and reservation levels, arbitrarily and subjectively defined by the negotiator. They may be defined by offers in which the single-issue resolution levels either fall outside the negotiator's predefined feasible domain or are certain intermediate levels within them. As a result, there is a need to evaluate offers with resolution levels worse than reservation levels (under-bad ones) and those with resolution levels better than aspiration levels (over-good ones). Two different approaches to handling the extent to which an alternative may outperform the aspiration level (or underperform the reservation level) are proposed: (1) non-compensatory one that simply omits the single-issue surpluses over aspiration levels and shortages to reservation levels when determining the offer global score, and (2) compensatory one that replaces the notion of distances with positive deviations (over aspiration level) and negative deviations (below reservation level) that are incorporated into the formula used to determine the offer globals score. Some similar modifications to the TOPSIS procedure were previously proposed for the crisp and fuzzy TOPSIS method [18, 19]; however, in this paper, we extended them by introducing a general framework for the evaluation of offers outside the negotiation space and discussing the impact of four normalization formulas for evaluation and rank-ordering of negotiation offers.

The paper consists of four more sections. Section 2 outlines problems with classic TOPSIS and its implementation to scoring negotiation offers when they fall beyond the previously defined feasible negotiation space and proposes two approaches to handle them: compensatory and non-compensatory. In Sect. 3, the extended TOPSIS algorithm for the evaluation of over-good and under-bad options in negotiations is presented. Section 4 presents an example of using enhanced TOPSIS to evaluate the negotiation template. To emphasize its advantages, the simulation investigating the instability of the scoring system for new offers of the classic TOPSIS is also shown. The paper finishes with conclusions.

2 Problems with Classic TOPSIS and Its Implementation to Scoring Negotiation Offers

The classic TOPSIS procedure, originally proposed by Hwang and Yoon [8], incorporates fundamental notions of reference alternatives utilized for ordering objects in multivariate statistics and was introduced earlier by Hellwig [7]. TOPSIS is based upon a classic definition of an MCDM problem, wherein $A = \{A_1, \ldots, A_m\}$ constitutes a finite set of m alternatives (representing negotiator offers), and $C = \{C_1, \ldots, C_n\}$ forms a finite set of n criteria (representing negotiation issues). Each alternative is then expressed as a vector $A_i = [x_{i1}, \ldots, x_{in}]$, where $x_{ij} \in \mathcal{R}$ denotes the resolution level of offer A_i for criterion C_j. All alternatives, along with their resolution levels for each criterion, collectively form a decision matrix $D = \left[x_{ij}\right]_{m \times n}$. Within set C, two subsets of benefit (I) and cost (J) criteria are defined, such that $I \cup J = C, I \cap J = \varnothing$. The decision maker (negotiator) is tasked with assigning criteria weights $w = [w_1, \ldots, w_n]$, descibing the issue's importance, subject to the constraint $\sum_{i=1}^{n} w_i = 1$. From D, reference positive ideal solution $A^+ = [x_1^+, \ldots, x_n^+]$ and negative ideal solution $A^- = [x_1^-, \ldots, x_n^-]$ are derived, being max or min values in columns of D, depending on whether it refers to benefit or cost criterion.

With the input data defined above, TOPSIS then requires determining the weighted normalized decision matrix as $\tilde{D} = \left[\tilde{x}_{ij}\right]_{m \times n}$, where $\tilde{x}_{\text{ij}} = w_j \bar{x}_{ij}$, and $\bar{x}_{ij}$ are obtained from x_{ij} using a selected normalization procedure. Also, the normalized weighted reference positive ideal solution $\tilde{A}^+ = [\tilde{x}_1^+, \ldots, \tilde{x}_n^+]$ and normalized weighted negative ideal solution $\tilde{A}^- = [\tilde{x}_1^-, \ldots, \tilde{x}_n^-]$ are derived. For each alternative A_i, the distances d_i^+ and d_i^- to A^+ and A^- are determined using selected metrics based on their normalized and weighted values. Finally, a global score for each alternative is determined as a relative closeness to the reference alternatives $T_i = \frac{d_i^-}{d_i^- + d_i^+}$. The values of T_i are nicely scaled to [0;1]-range, allowing an intuitive interpretation of the quality of alternatives, with the best (ideal) alternative scoring 1 and the worst one – 0.

If the negotiation scoring system is constructed using classic TOPSIS, it will be unable to score offers that consist of resolution levels higher than x_j^+ or lower than x_j^-

since they might fall outside the [0;1] rating scale. It implies that offers better than A^+ or worse than A^- cannot be submitted to the negotiation table. This scenario is feasible only when parties collaboratively define a workable negotiation space that encompasses solutions meeting all one-sided aspirations or expectations toward maximum counterparts' concessions. Regrettably, in practice, parties seldom engage in joint pre-negotiation preparation [20, 22], potentially leading them to propose offers beyond the set of feasible solutions defined by one or all parties.

Making TOPSIS capable of evaluating such offers requires some modifications in the scoring procedure. This adaptation must consider that hard constraints regarding the form of the negotiation space may be unknown to the negotiator (due to a lack of joint preparation). As a result, the reference alternatives on which the assessment system relies cannot be defined based on it. Therefore, the modification we propose substitutes the concept of ideal and anti-ideal solutions with negotiation-specific notions of aspiration and reservation levels. It is assumed that the latter are independent of the decision matrix (negotiation space) but are entirely subjectively defined by the negotiator based on their preferences. We denote aspiration (A^{asp}) and reservation (A^{res}) levels as.

$$A^{asp} = \left[x_1^{asp}, \ldots, x_n^{asp}\right], \tag{1}$$

$$A^{res} = [x_1^{res}, \ldots, x_n^{res}] \tag{2}$$

where x_j^{asp} (x_j^{res}) is the aspiration (reservation) level for the j th criterion.

Furthermore, we assume that negotiators acknowledge that various offers may occur during the negotiation process, including ones consisting of options worse than assumed reservation levels or exceeding aspiration levels. We call such options under-bad or over-good ones and formally define them in the following way.

Definition 1
An option x_{ij} is over-good if

$$\begin{cases} x_{ij} > x_j^{\text{asp}} & \text{for benefit criterion} \\ x_{ij} < x_j^{\text{asp}} & \text{for cost criterion} \end{cases}. \tag{3}$$

Definition 2
An option x_{ij} is under-bad if

$$\begin{cases} x_{ij} < x_j^{res} & \text{for benefit criterion} \\ x_{ij} > x_j^{res} & \text{for cost criterion} \end{cases}. \tag{4}$$

Evaluating over-good or under-bad options in the modified TOPSIS algorithm is not straightforward. It can be approached in two distinct ways, depending on how the negotiator views the possibility of incorporating these options in determining trade-offs between criteria.

The most primitive way is to ignore that over-good or under-bad options fall outside the feasible range of resolution levels of the issue under consideration. This approach is based on a specific philosophy of looking at the problem by the negotiator, who assumes that the prepared scoring system considers all the relevant solutions, and even though some others emerge, they are considered equivalent to those from the set of feasible solutions defined beforehand. Consequently, if a particular option exceeds the aspiration level, it will not be assigned a score higher than this aspiration level (considered better). If it is worse than the reservation level, it will not be scored lower than this reservation level (considered worse). Technically, performances x_{ij} better than x_j^{asp} will be neglected in compensating for the potentially worse-than-best performances of A_i on other issues but also performances x_{ij} worse than x_j^{res} will not add to shortages of worse-than-best performance of A_i on this issue. For this reason, we name this approach ***non-compensatory***. In Fig. 1a, we visualize the idea of a non-compensatory analysis of the over-good option. An alternative A_i is considered to fall outside feasible space P, i.e., is over-good for criterion C_1 (exceeds it of 3 units). Non-compensatory evaluation of A_i requires considering it as A'_i. Its surpluses over x_1^{asp} are omitted, as the negotiator does not attach any value to them. Hence, the distance between A_i and A^{asp} is measured by $d_i'^+$ and between between A_i and A^{res} – by $d_i'^-$.

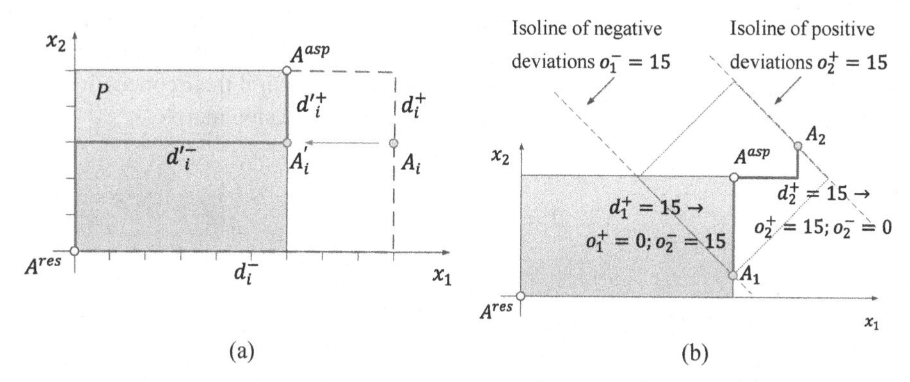

Fig. 1. Evaluation of over-good option according to non-compensatory (a) and compensatory (b) approach in enhanced TOPSIS

However, some negotiators may wish to consider the fact that A_i surpasses the aspirational levels on certain issues and includes these excesses in the trade-off analysis (***compensatory*** approach). In such a scenario, the concept of measuring distances proves insufficient. From Fig. 1a, one can easily conclude that A_i may turn out to be inferior to A'_i because it is more distant from A^{asp} than A'_i (indeed, $T_{A_i} = 0.71$, while $T_{A'_i} = 0.82$ when Manhattan distance is used). The concept of distance does not allow us to adequately address the situation in which some distance may be treated as a positive phenomenon for the negotiator.

Thus, when compensation is to be addressed, we replace the concept of distance with the notion of deviations. It allows for distinguishing between positively perceived distance (desired from the negotiator's preference perspective, describing how much better an option is) - termed *positive deviation* (o^+), and negatively perceived distance (undesirable, indicating how much an option falls short compared to the reference option) -

termed negative deviation (o^-). It is visualized in Fig. 1b when two offers are considered, both having the same distance to A^{asp} ($d_1^+ = d_2^+$) but the distance of the former has a negative connotation for the negotiator, while the latter has a positive one. Thus $o_1^+ \neq o_2^+$ and $o_1^- \neq o_2^-$. Positive deviations should decrease the offer's separation measure to A^{asp} and increase the separation measure to A^{res} (opposed should be an effect of negative deviations). Computational details of how deviations affect separation measures are discussed in the following section.

3 Enhanced TOPSIS Procedure for Evaluation of Over-Good and Under-Bad Options in Negotiations

3.1 Enhanced TOPSIS Routine

The enhanced TOPSIS procedure that accommodates the negotiator's individual aspiration and reservation levels and his various perceptions of compensation consists of the following six steps:

Step 1. Designing a negotiation template as the set of negotiation issues C_j ($j = 1, 2, \ldots n$) and the set of some exemplary offers $A_i (i = 1, \ldots, m)$, subjectively selected by the negotiator, e.g., the ones the negotiator considers to build their concession path for the negotiation strategy [see, e.g., 17], and building a decision matrix

$$D = \begin{bmatrix} x_{11} & \cdots & x_{1n} \\ \vdots & \ddots & \vdots \\ x_{m1} & \cdots & x_{mn} \end{bmatrix} \tag{5}$$

that specifies the numerical representation of resolution levels x_{ij} of each alternative A_i with respect to each criterion C_j.

Step 2. Identifying preferences in the form of the criteria weights w and the aspiration and reservation levels (A^{asp} and A^{res}). The weights are defined as a vector

$$w = [w_1, \ldots, w_n], \tag{6}$$

where $w_j > 0$ ($j = 1, \ldots, n$) is the importance of criterion C_j in the global evaluation of any negotiation offer and $\sum_{i=1}^{n} w_i = 1$. We assume that the weights are provided subjectively by the negotiator and remain stable during the negotiation process. Aspiration and reservation levels are defined using formulas (1) and (2).

Step 3. Determining the weighted normalized decision matrix

$$\tilde{D} = \begin{bmatrix} \tilde{x}_{11} & \cdots & \tilde{x}_{1n} \\ \vdots & \ddots & \vdots \\ \tilde{x}_{m1} & \cdots & \tilde{x}_{mn} \end{bmatrix} \tag{7}$$

where $\tilde{x}_{ij} = w_j \bar{x}_{ij}$ represent the weighted normalized performances of all offers from D. Here, $\bar{x}_{ij}$ are the normalized performances obtained from the selected normalization formula (see Sect. 3.2 for examples) applied to the columns of matrix D.

Step 4. Determining the weighted normalized aspiration and reservation levels

$$\tilde{A}^{asp} = \left[\tilde{x}_1^{asp}, \ldots, \tilde{x}_n^{asp}\right], \tag{8}$$

$$\tilde{A}^{res} = \left[\tilde{x}_1^{res}, \ldots, \tilde{x}_n^{res}\right], \tag{9}$$

where $\tilde{x}_j^{asp} = w_j \overline{x}_j^{asp}$, $\tilde{x}_j^{res} = w_j \overline{x}_j^{res}$, and $\overline{x}_j^{asp}(\overline{x}_j^{res})$ are normalized aspiration and reservation levels obtained from the same formulas used in step 3.

Step 5. Determining separation measures for each alternative A_i with respect to aspiration A^{asp} and reservation A^{res}.

In the ***non-compensatory*** approach, separation measures utilize the notion of distances and the performance values from $\tilde{D}$, which are modified accordingly, depending on whether they represent benefit or cost issues. For benefit ones, modified performances are determined as

$$\overline{\overline{x}}_{ij} = \begin{cases} \tilde{x}_i^{asp} \text{ if } \tilde{x}_{ij} > \tilde{x}_i^{asp} \\ \tilde{x}_i^{res} \text{ if } \tilde{x}_{ij} < \tilde{x}_i^{res} \\ \tilde{x}_{ij} \quad \text{otherwise} \end{cases}, \tag{10}$$

while for the cost ones as

$$\overline{\overline{x}}_{ij} = \begin{cases} \tilde{x}_i^{asp} \text{ if } \tilde{x}_{ij} < \tilde{x}_i^{asp} \\ \tilde{x}_i^{res} \text{ if } \tilde{x}_{ij} > \tilde{x}_i^{res} \\ \tilde{x}_{ij} \quad \text{otherwise} \end{cases}. \tag{11}$$

They are used to determine the separation values to A^{asp} and A^{res} employing the classical formula of Minkowski distance:

$$s_i^{asp} = \sqrt[p]{\left|\overline{\overline{x}}_{ij} - \tilde{x}_i^{asp}\right|^p}, \tag{12}$$

$$s_i^{res} = \sqrt[p]{\left|\overline{\overline{x}}_{ij} - \tilde{x}_i^{res}\right|^p}. \tag{13}$$

In the ***compensatory*** approach, the separation values are calculated using notions of positive and negative deviations in the following way

$$s_i^{asp} = q_i^{+} \sqrt[p]{\left|\sum_{j=1}^{n} q_{ij}^{+} \left|\tilde{x}_{ij} - \tilde{x}_i^{asp}\right|^p\right|}, \tag{14}$$

$$s_i^{res} = q_i^{-} \sqrt[p]{\left|\sum_{j=1}^{n} q_{ij}^{-} \left|\tilde{x}_{ij} - \tilde{x}_i^{res}\right|^p\right|}, \tag{15}$$

where q_{ij}^{+} (q_{ij}^{-}) are indicators of the direction of single-criterion deviation from A^{asp} (A^{res}) determined from the formulas:

$$q_{ij}^{+} = \begin{cases} 1 \text{ if } \tilde{x}_i^{asp} > \tilde{x}_{ij} \text{ for } C_j \in I \text{ or } \tilde{x}_i^{asp} < \tilde{x}_{ij} \text{ for } C_j \in J \\ -1 \text{ if } \tilde{x}_i^{asp} \le \tilde{x}_{ij} \text{ for } C_j \in I \text{ or } \tilde{x}_i^{asp} \ge \tilde{x}_{ij} \text{ for } C_j \in J \end{cases}, \tag{16}$$

$$q_{ij}^{-} = \begin{cases} 1 & \text{if } \tilde{x}_i^{res} < \tilde{x}_{ij} \text{ for } C_j \in I \text{ or } \tilde{x}_i^{res} > \tilde{x}_{ij} \text{ for } C_j \in J \\ -1 & \text{if } \tilde{x}_i^{res} \geq \tilde{x}_{ij} \text{ for } C_j \in I \text{ or } \tilde{x}_i^{res} \leq \tilde{x}_{ij} \text{ for } C_j \in J \end{cases}, \quad (17)$$

and q_i^+ (q_i^-) are indicators of the direction of aggregate deviation from A^{asp} (A^{res}):

$$q_i^+ = \begin{cases} 1 & \text{if } \sum_{j=1}^{n} q_{ij}^+ |\tilde{x}_{ij} - \tilde{x}_i^{asp}|^p > 0 \\ -1 & \text{otherwise} \end{cases}, \quad (18)$$

$$q_i^- = \begin{cases} 1 & \text{if } \sum_{j=1}^{n} q_{ij}^- |\tilde{x}_{ij} - \tilde{x}_i^{res}|^p > 0 \\ -1 & \text{otherwise} \end{cases}. \quad (19)$$

Step 6. Determining global separation measure using a classic TOPSIS idea

$$T_i = \frac{s_i^-}{s_i^- + s_i^+}. \quad (20)$$

3.2 The Normalization Formulas Used in TOPSIS

Normalization is a crucial step in most MCDM techniques, as it transforms input data, which may be measured in different units, into numerical and comparable data. Several normalization techniques have been developed in the literature [1, 4, 5, 8, 9], and their impact on the ranking of alternatives obtained by MCDM methods has been studied by many authors [4, 13, 16, 27]. Below, see Table 1, we present four normalization procedures usually employed in TOPSIS analyses. However, in this paper, we will test the enhanced TOPSIS technique behavior using only max-min (N1) and sum (N3) normalization approaches.

Table 1. Examples of normalization formulas used in TOPSIS

Normalization	Formula
linear scale transformation, max-min method (N1)	$\overline{x}_{ij} = \frac{x_{ij} - min(x_j^{res}, x_j^{asp})}{max(x_j^{res}, x_j^{asp}) - min(x_j^{res}, x_j^{asp})}$
linear scale transformation, max method (N2)	$\overline{x}_{ij} = \frac{x_{ij}}{\max(x_j^{res}, x_j^{asp})}$
linear scale transformation, sum method (N3)	$\overline{x}_{ij} = \frac{x_{ij}}{\sum_{i=1}^{m} x_{ij}}$
vector normalization (N4)	$\overline{x}_{ij} = \frac{x_{ij}}{\sqrt{\sum_{i=1}^{m} (x_{ij})^2}}$

Note: Similarly, the aspiration (reservation) levels are normalized, with minor modifications requiring $\overline{x}_{ij} = \overline{x}_j^{asp}$ ($\overline{x}_{ij} = \overline{x}_j^{res}$) and $x_{ij} = x_j^{asp}$ ($x_{ij} = x_j^{res}$).

3.3 Enhanced TOPSIS and New Offers in Actual Negotiations

It is worth noting that the proposed enhanced TOPSIS allows, as intended, to evaluate alternatives in the pre-negotiation phase (defined in the form of matrix D) with consideration of individual reference levels defined independently by the negotiator. At this stage, the determined scoring system takes into account the potential alternatives with over-good and under-bad options. However, this scoring system can also be used in the actual negotiation phase to assess new negotiation offers, even if they consist of options beyond the domain defined by the values in matrix D. The evaluation of such an offer follows the scheme of assessing aspiration and reservation levels as included in the algorithm from Sect. 3.1. The normalization of options for such offers is carried out using the formulas proposed in Sect. 3.2 (Table 1), providing that $\overline{x}_{ij} = \overline{x}_j^{new}$ and $x_{ij} = x_j^{new}$. The advantage of this solution is that the scoring system remains stable, i.e., regardless of the number of new offers subjected to assessment, evaluations of offers from D remain unchanged.

4 Using Enhanced TOPSIS to Evaluate Negotiation Template

4.1 Case Description

This section presents the enhanced TOPSIS mechanism for evaluating a negotiation template. We also draw attention to the differences in the obtained scoring systems resulting from various compensatory approaches and normalization concepts utilized. We use an example of a negotiation scenario in which an artist (Fado) and a broadcasting company (Mosico) negotiate the terms of a contract that includes four issues: C_1 signing bonus (kUSD), C_2 royalties for CDs (%), C_3 no. of concerts and C_4 no. of songs [19].

Let us assume that we build a scoring system with enhanced TOPSIS for Fado who, following step 1, has defined a set of alternative contract proposals, i.e.,

$$D = \begin{matrix} A_1 \\ A_2 \\ A_3 \\ A_4 \\ A_5 \\ A_6 \end{matrix} \begin{bmatrix} 325, 3.5, 5, 12 \\ 150, 2.5, 15, 20 \\ 300, 2.5, 5, 16 \\ 325, 1.0, 8, 20 \\ 200, 5.0, 4, 14 \\ 220, 2.0, 5, 17 \end{bmatrix}.$$

Within step 2, she specified the benefit and cost issues ($I = \{C_1, C_2\}, J = \{C_3, C_4\}$), as well as their weights $w = [0.4; 0.3; 0.2; 0.1]$. Additionally, she has established aspiration and reservation levels as $A^{asp} = [325; 3.5; 5; 12]$ and $A^{res} = [100; 1.75; 10; 18]$. Consequently, we see that A_1 is equivalent to aspiration level, A_2 and A_4 consist of an under-bad option for C_3 and C_4, and C_2 and C_4, respectively, while offer A_5 of an over-good option for C_2 and for C_3. Let us note that ideal positive and ideal negative solutions determined as max and min values from matrix D are as follows: $A^+ = [325; 5; 4; 12]$ and $A^- = [150; 1; 15; 20]$.

4.2 Comparisons of Compensatory and Non-Compensatory Approaches

Various normalization procedures or distance measures may impact the results of the aggregation of single-criterion performances [1, 5, 15]. Thus, we may expect the issues will also play a role in differentiating results from non-compensatory and compensatory approaches used for scoring the negotiation space D. Table 2 presents the ratings and rankings of offers from D determined by enhanced TOPSIS with A^{asp} and A^{res} and for classic TOPSIS with ideal positive A^{+} and ideal negative A^{-}. Two normalization formulas were implemented, N1 and N3, and the Manhattan distance measure.

Table 2. The evaluation of negotiation offers for normalization formulas N1, N3, and p = 1

Offers	**N1**			**N3**		
	non-com	com	classic	non-com	com	classic
A1	1.000(1)	1.000(2)	0.869(1)	1.000(1)	1.000(1)	0.821(1)
A2	0.217(6)	**−0.016(6)**	0.113(6)	0.222(6)	**0.008(6)**	0.152(6)
A3	0.717(3)	0.717(3)	0.687(3)	0.762(2)	0.762(3)	0.661(3)
A4	0.480(4)	**0.318(5)**	0.527(4)	0.569(4)	**0.439(5)**	0.443(5)
A5	0.744(2)	**1.042(1)**	0.689(2)	0.711(3)	**0.976(2)**	0.805(2)
A6	0.473(5)	0.473(4)	0.454(5)	0.504(5)	0.504(4)	0.487(4)

The results of the comparison of scoring systems obtained through different TOPSIS approaches are summarized as follows:

- The classic approach yields slightly different rankings depending on the normalization formula. The differences concern the assessment of offers $A4$ and $A6$, which take positions 4, 5 for N1, and 5, 4 for N3, respectively.
- Offer $A1$ is an aspiration level and reference point for evaluating other offers under both compensatory and non-compensatory approaches. Therefore, in both normalizations, it obtains a rating of 1. It is also the best in the classic TOPSIS but with ratings <1, resulting from the positive ideal solution being different from the individual aspiration level.
- Offer $A2$ has two under-bad options. Additionally, in the first criterion, it achieves the lowest value of all offers in D and only slightly exceeds the reservation level in C_2. Therefore, regardless of the normalization method, it receives the lowest rating in each approach. In the compensatory approach with N1, it falls below 0, which clearly indicates it is inferior to the reservation level. Unfortunately, this does not hold for N3.
- Offer $A3$ does not include over-good or under-bad options. Therefore, the assessment for compensatory and non-compensatory approaches is the same. It differs from classic TOPSIS, as the latter uses different reference alternatives to evaluate all offers from D.
- Offer $A4$ contains two under-bad options, resulting in a lower rating in the compensatory approach. In the non-compensatory approach, offer $A4$ is better than offer $A6$,

while in the compensatory approach, it is the opposite. Offer $A6$ does not include over-good or over-bad options. Therefore, its rating does not depend on the adopted approach (compensatory or non-compensatory). In the classic TOPSIS approach, offers $A4$ and $A6$ switch places (4 or 5) depending on the normalization formula.

- Offer $A5$ contains two over-good options, resulting in a higher rating for the compensatory approach compared to the non-compensatory one. In the non-compensatory approach, offer $A5$ was rated as 2 or 3 depending on the normalization method. In the classic approach, it was rated as second. However, in the compensatory approach, it moved up to the first position, surpassing even the aspiration level for normalization method N1. For the N3 normalization method, it also moved up one position to second place, but with a lower score than offer $A1$.
- Similar to $A4$, A6 does not include over-good or under-bad options, but its position varies and depends on the normalization formula and compensation approach. In the classic TOPSIS approach, $A6$ ranks 5th for N1 and 4th for N3. In enhanced TOPSIS, it occupies the fifth position for the non-compensatory approach and the fourth position for the compensatory approach. It is better than $A4$, which, due to the under-bad option, lost a lot in compensation.

4.3 Stability of Scoring System for New Offers

To emphasize the advantage of enhanced compensatory TOPSIS, we compared it to classic TOPSIS with corresponding max-min scaling (both adopting N1) in terms of the stability of ratings (and rankings) for offers from set D, evaluated during pre-negotiations, given the new offers with under-bad options may potentially be submitted by the counterpart. We assumed that Mosico thoroughly defined the feasible negotiation space in pre-negotiations in the form of 240 offers spanning the permissible domains of values for all issues: C_1 ={125, 150, 200}, C_2 ={1.5, 2, 2.5, 3}, C_3 ={5, 6, 7, 8}, and C_4 ={11, 12, 13, 14, 15}. The classic TOPSIS mechanism with max-min scaling ensures a clear ranking of these offers with reference offers $A^+ = [125, 1.5, 8, 15]$ and $T_A^+ = 1$, and $A^- = [200, 3, 5, 11]$ and $T_A^- = 0$. Enhanced TOPSIS ensures the same if we assume $A^+ = A^{asp}$ and $A^- = A^{res}$. However, we assumed that Mosico underestimated the common feasible negotiation space (by approximately 20% of the values of A^-), and during the negotiation opening, her counterpart submitted an offer worse than A^-. We conducted a simulation of such counteroffers, assuming equal probability of choosing options within each issue C_j in the range of $\pm 20\%$ of A_j^-. We then examined how the ranking and rating of offers from set D, determined using classic TOPSIS, changed due to the necessity of recalculating the ratings of all 241 offers using classic TOPSIS algorithms (while results from enhanced TOPSIS do not change in such a situation). We determined three measures indicating the degree of changes in the results: fraction of offers with a change in ranking, average ranking change, and average rating difference. The results are presented in Table 3.

It is clearly evident that the new offer A^{new} from outside the feasible negotiation space changes the evaluations of offers from D since recalculations are required by classic TOPSIS for $D \cup A^{new}$. On average, the rating of offers from set D changes by 0.073. Considering that the rating space is normalized to the range [0; 1], the average differences amount to 7.3% of the rating space! These differences escalate when the negotiator's imprecision in defining the feasible negotiation space increases, and the new offers contain more under-bad options. They are less noticeable when analyzed at the ranking level. Still, on average, each offer differs in eleven ranks after adding A^{new}, and nearly 95% of offers change their ranking position.

Table 3. The evaluation of negotiation offers for normalization formulas N1, N3, and p = 1

Number of under-bad options in A^{new}	Fraction of offers that changed rank	Average change in rank of each offer	Average change in rating of each offer
1	0.946	11.6	0.042
2	0.949	12.0	0.058
3	0.949	12.1	0.081
4	0.949	10.0	0.112
Average	0.948	11.5	0.073

It is worth noting that after receiving an under-bad offer in the next round, Mosico will submit a counteroffer that exceeds her aspiration levels (to change the anchors and restore the balance between them). Consequently, the evaluations of offers from D will change again, making her probably more confused. When the scoring system is unstable, the negotiator may lose interest in using it since it does not allow for drawing reliable conclusions regarding offer values.

5 Conclusion

This paper presents the concept of evaluating negotiation offers and building a negotiation scoring system. It is based on aspiration and reservation levels determined individually, which define the feasible negotiation space. The procedure of enhanced TOPSIS enables the assessment of offers outside the feasible negotiation space using a compensatory or non-compensatory approach to evaluate over-good and under-bad options. Applying the enhanced TOPSIS method to build an offer evaluation system simplifies the process of building a negotiation offer system, as it eliminates the cumbersome process of assigning scoring points to all negotiation issues and options by negotiators individually. On the other hand, it allows for imparting individual preferences in terms of aspiration and reservation levels to score the feasible negotiation space by implementing the notion of distances or deviations. Additionally, it allows easy identification of global over-good or under-bad options as they obtain scores higher than aspiration (1) or lower than reservation level (0). Finally, it ensures the stability of ratings of predefined negotiation space no matter how many new offers from outside this space are submitted later during the negotiation process, which is vital from the viewpoint of univocal interpretation of negotiation history and dynamics.

It should be noted that compensatory and non-compensatory approaches cannot be used interchangeably, as they are based on a different philosophy of perceiving surpluses and deficiencies in the values of certain options and including them in trade-off analysis. It is essential to emphasize that the chosen normalization method can significantly influence the value of offers. Therefore, the negotiator must determine which compensation approach, scaling technique, and the notion of distance aligns with their perspective on the negotiation problem and its context. Establishing the best mix of these technical issues for a particular negotiation context will be the subject of our future work.

Acknowledgment. The contribution of Ewa Roszkowska to this research was financed by the grant WZ/WI-IIT/2/22 from the Bialystok University of Technology, while the contribution of Tomasz Wachowicz was supported by the "Regional Initiative of Excellence" program, funded by the Polish Ministry of Science and Higher Education.

References

1. Acuña-Soto, C., et al.: Normalization in TOPSIS-based approaches with data of different nature: application to the ranking of mathematical videos. Ann. Oper. Res. **296**(1), 541–569 (2021)
2. Brzostowski, J., et al.: Supporting negotiation by multi-criteria decision-making methods. Optimum-Studia Ekonomiczne. **5**(59), 59 (2012)
3. Brzostowski, J., et al.: Using an analytic hierarchy process to develop a scoring system for a set of continuous feasible alternatives in negotiation. Oper. Rese. Decis. **22**(4), 21–40 (2012)
4. Çelen, A.: Comparative analysis of normalization procedures in TOPSIS method: with an application to Turkish deposit banking market. Informatica **25**(2), 185–208 (2014)
5. Chakraborty, S., Yeh, C.-H.: A simulation comparison of normalization procedures for TOPSIS. In: Proceedings of the 2009 International Conference on Computers and Industrial Engineering (CIE39). pp. 1815–1820. IEEE, Institute of Electrical and Electronics Engineers (2009)
6. Górecka, D., et al.: The MARS approach in the verbal and holistic evaluation of the negotiation template. Group Decis. Negot. **25**(6), 1097–1136 (2016). https://doi.org/10.1007/s10726-016-9475-9
7. Hellwig, Z.: Procedure of evaluating high level manpower data and typology of countries by means of the taxonomic method (in Polish). Statist. Rev. **15**, 308–327 (1968)
8. Hwang, C.-L., Yoon, K.: Methods for multiple attribute decision making. In: Hwang, C.-L., Yoon, K. (eds.) Multiple Attribute Decision Making, pp. 58–191. Springer, Berlin, Heidelberg (1981). https://doi.org/10.1007/978-3-642-48318-9_3
9. Jahan, A., Edwards, K.L.: A state-of-the-art survey on the influence of normalization techniques in ranking: Improving the materials selection process in engineering design. Mater. Des. (1980–2015) **65**, 335–342 (2015)
10. Jarke, M., et al.: MEDIATOR: towards a negotiation support system. Eur. J. Oper. Res. **31**(3), 314–334 (1987)
11. Kersten, G.E., Noronha, S.J.: WWW-based negotiation support: design, implementation, and use. Decis. Support Syst. **25**(2), 135–154 (1999)
12. Kersten, G.E., Szapiro, T.: Generalized approach to modeling negotiations. Eur. J. Oper. Res. **26**(1), 142–149 (1986). https://doi.org/10.1016/0377-2217(86)90166-9
13. Milani, A.S., et al.: The effect of normalization norms in multiple attribute decision making models: a case study in gear material selection. Struct. Multidiscip. Optim. **29**(4), 312–318 (2005)

14. Mustajoki, J., Hämäläinen, R.P.: Web-hipre: global decision support by value tree and AHP analysis. INFOR: Inf. Syst. Oper. Res. **38**(3), 208–220 (2000). https://doi.org/10.1080/03155986.2000.11732409
15. Palczewski, K., Sałabun, W.: Influence of various normalization methods in PROMETHEE II: an empirical study on the selection of the airport location. Procedia Comput. Sci. **159**, 2051–2060 (2019). https://doi.org/10.1016/j.procs.2019.09.378
16. Pavličić, D.: Normalization affects the results of MADM methods. Yugoslav J. Oper. Res. **11**(2), 251–265 (2001)
17. Raiffa, H., et al.: Negotiation Analysis: The Science and Art of Collaborative Decision Making. Harvard University Press, Cambridge (2002)
18. Roszkowska, E., Wachowicz, T.: Application of fuzzy TOPSIS to scoring the negotiation offers in ill-structured negotiation problems. Eur. J. Oper. Res. **242**(3), 920–932 (2015). https://doi.org/10.1016/j.ejor.2014.10.050
19. Roszkowska, E., Wachowicz, T.: Ocena ofert negocjacyjnych spoza dopuszczalnej przestrzeni negocjacyjnej. Prace Naukowe Uniwersytetu Ekonomicznego we Wrocławiu. **385**, 202–209 (2015)
20. Stein, J.G.: Pre-negotiation in the Arab-Israeli Conflict: the paradoxes of success and failure. Int. J. **44**(2), 410–441 (1989). https://doi.org/10.1177/002070208904400207
21. Thiessen, E.M., Soberg, A.: Smartsettle described with the montreal taxonomy. Group Decis. Negot. **12**(2), 165 (2003)
22. Tomlin, B.W.: The stages of pre-negotiation: the decision to negotiate North American free trade. Int. J. **44**(2), 254–279 (1989)
23. Wachowicz, T.: Negotiation Template Evaluation with Calibrated ELECTRE-TRI Method, pp. 232–238. Group Decision and Negotiations. The Center for Collaboration Science, University of Nebraska at Omaha (2010)
24. Wachowicz, T., Błaszczyk, P.: TOPSIS based approach to scoring negotiating offers in negotiation support systems. Group Decis. Negot. **22**(6), 1021–1050 (2013)
25. Wachowicz, T., Roszkowska, E.: Can holistic declaration of preferences improve a negotiation offer scoring system? Eur. J. Oper. Res. **299**(3), 1018–1032 (2022). https://doi.org/10.1016/j.ejor.2021.10.008
26. Wachowicz, T., Roszkowska, E.: Holistic preferences and prenegotiation preparation. In: Kilgour, D.M., Eden, C. (eds.) Handbook of Group Decision and Negotiation, pp. 255–289. Springer, Cham (2021)
27. Zavadskas, E.K., et al.: Evaluation of ranking accuracy in multi-criteria decisions. Informatica **17**(4), 601–618 (2006)

A Stratified Fuzzy Group Best Worst Decision-Making Framework

Yanlin Li[1], Y. P. Tsang[1](✉), C. K. M. Lee[1], and Yipu Yao[2]

[1] Department of Industrial and Systems Engineering, The Hong Kong Polytechnic University, Hong Kong Special Administrative Region, China
yungpo.tsang@polyu.edu.hk

[2] Durham Business School, Durham University, Durham, UK

Abstract. The importance of multi-criteria group decision-making techniques in today's complex decision-making environment is undeniable. In complex decision problems, the determination of criteria weights is often subject to both DM-related uncertainty and external event-related uncertainty. The Fuzzy Best-Worst Method (FBWM), as a widely applied MCDM technique based on pairwise comparisons under the fuzzy environment, has been recognized for its systematic expert data collection process, reduced pairwise comparison steps, and high result consistency. Considering the impact of future events on current decision criteria is also necessary in practice. The concept of stratification (CST) provides a fresh perspective on the stratified decision-making process in the MCDM domain. However, the integration of CST with MCDM methods is still in its infancy. Specifically, there is a lack of exploration in the implementation process of FBWM under CST in a group decision-making environment. To address this gap, this study proposes a stratified fuzzy best-worst group decision-making framework. This study aims to promote human-centric decision-making by incorporating advanced decision-making techniques to facilitate efficient decision implementation. An illustrative numerical example is provided in this paper to demonstrate the applicability of the proposed framework.

Keywords: Group Decision Making · Concept of Stratification · Stratified Fuzzy Best-Worst Group Method · Human-centric Decision-Making

1 Introduction

The growing complexity of modern socioeconomic environments makes it difficult for an individual decision maker (DM) to consider all aspects of a decision problem (Kim & Ahn, 1999). Many complicated decision problems such as supplier selection, product design and development, resource allocation, and market segmentation, which requires consider various evaluation criteria, are often made by the help of a group of advisors and experts. However, moving from individual multi-criteria decision-making (MCDM) to a group MCDM adds to the complexity of the processing and analysis (Hafezalkotob & Hafezalkotob, 2017). In addition, with the development of globalization in business, the

M. Campos Ferreira et al. (Eds.): GDN 2024, LNBIP 509, pp. 65–76, 2024.
https://doi.org/10.1007/978-3-031-59373-4_6

impact of future events becomes increasingly important for current business decision-making (Ecer & Torkayesh, 2022). For instance, with the increasing worldwide concerns about global warming, more and more countries are taking actions related to it, such as Germany's ESG legislation and the European CBAM (carbon taxes). When a manufacturer starts selecting potential suppliers for its global business, the importance of evaluation criteria related to environmental impact may increase, considering the legalization trend in importing countries. In this context, more and more DMs claim the necessity of considering the impact of future events in the group decision-making (GDM) process.

According to Asadabadi and Zwikael (2021), the incorporation of future events into the decision-making process is expected to result in more reasonable decision outcomes. The stratified MCDM (S-MCDM) method, proposed by Asadabadi (2018), is based on the concept of stratification (CST) introduced by Zadeh (2016). The S-MCDM method has been developed as an effective approach to addressing uncertainty by considering all the potential scenarios that may occur in the near future (Zadeh, 2016; Asadabadi, 2018). The central idea of S-MCDM is to account for fluctuations in the weights of decision criteria, in a manner that resembles the way changes occur in the human brain, thereby addressing the limitations of traditional MCDM methods in handling uncertainty. The stratified decision-making strategy in S-MCDM mirrors the cognitive processes of the human brain. When the human brain deliberates on several criteria to arrive at the optimal decision, it factors in a multitude of favorable and unfavorable scenarios that may materialize (Asadabadi, 2018). For instance, when deciding to purchase a property, DMs may consider various future events, such as the birth of a child, friends' visit, or a new job offer. These uncertainties can affect the weights of criteria such as the location, size, and price of the property. As the importance of the evaluation criteria changes, the ranking of the alternative options may also change, ultimately impacting the final decision outcome. It has been noted that contemplating the decision-making environment and envisaging potential situations can bolster the resilience of the ultimate decision outcome (Asadabadi, 2018).

Broadly speaking, uncertainty in the determination of criteria weights in decision-making processes is comprised of both DM-related uncertainty and external event-related uncertainty (Asadabadi et al., 2023). While both fuzzy sets and the CST were introduced by Zadeh (1965, 2016), they are fundamentally different concepts. Fuzzy sets address the absence of sharply defined criteria for class membership, whereas CST deals with the stochastic nature of problems (Asadabadi & Zwikael, 2021). In the complex real-life GDM scenarios, a flexible and robust decision framework that considers both the inherent uncertainty of human decision-making and the uncertainty associated with the occurrence of future events is lacking for implementation.

Present study contributes to the methodological innovation in the MCGDM domain. To the best of the authors' knowledge, this is the first study to formulate a subjective weight determination framework under SF-MCGDM scenarios. A structured best-worst preference-based group decision support framework incorporating the vagueness of human experts' preferences and the uncertainty of future events is developed. To enhance clarity, a simple numerical example is provided in this study to illustrate the deployment of the proposed decision framework.

The remainder of the paper is organized as follows: Sect. 2 provides a literature review related to CST in MCDM and fuzzy best-worst method in GDM. Section 3 presents the developed multi-criteria group decision-making framework. In Sect. 4, a numerical example is given. Finally, Sect. 5 concludes this article.

2 Literature Review

2.1 The Concept of Stratification in Multiple Criteria Decision

The potential of combining CST with MCDM methods has been recognized in the literature (Torkayesh et al., 2021; Ecer & Torkayesh, 2022). However, not all of the capabilities and components of CST are necessarily utilized in the MCDM domain. The integration of CST and MCDM typically depends on the core concept of stratifying the decision environment (Asadabadi, 2018). Owing to space constraints, this paper focuses on the relative concepts related to the application of CST in multiple criteria decision-making. Readers interested in more comprehensive knowledge of CST are encouraged to refer to the works of Zadeh (2016) and Asadabadi (2018). Here, some fundamental concepts of the decision-making process under CST are provided below.

States, strata, and target set in CST: Stratification categorizes a number of states which belong to different strata where one or more states are considered as the members of the target set. A state can be predefined by a domain expert. CST transitions through the states to reach a state in the target set. Figure 1 shows the target set, strata, and states of a conceptual CST (Zadeh, 2016; Asadabadi, 2018).

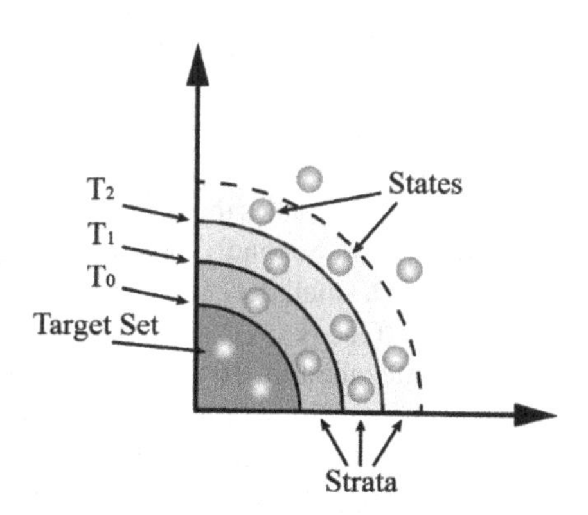

Fig. 1. Strata, states, and the target set of a conceptual CST.

Criterion weight in each state: Considering each situation as a state in CST, the weights for criteria in k th state are $W_k : \{w_{k1}, \cdots, w_{km}\}$. The criteria weights will vary depending on whether the current decision system remains unchanged or undergoes a transition to different states (Asadabadi, 2018).

State-transition probabilities: Assume that the decision system is at state 'k', the probability of transitioning to other state 'j' is P_{kj}. Note that P_{kk} denotes the probability of transitioning from state 'k' to state 'k', which is the likelihood that the current situation persists (Asadabadi, 2018).

Given that the S-MCDM concept remains relatively new, the existing literature on this topic remains limited in scope (Ecer & Torkayesh, 2022; Torkayesh et al., 2021). Consequently, there is considerable potential for exploring innovative integrations of MCDM techniques and CST, which takes into account events that may occur and influence the weights of criteria.

2.2 Fuzzy Best Worst Method Under Group Decision Scenario

In the field of discrete MCDM, there are a number of subjective weighting methods, such as Analytic Hierarchy Process (AHP), Best-Worst Method (BWM), Simple Multi-attribute Ranking Technique (SMART), and others. Compared to other similar MCDM methods, BWM offers advantages such as time savings for experts, ease of implementation, and more reliable and consistent results (Rezaei, 2015). However, in the original BWM, DMs are required to perform subjective comparisons by using the numerical scales from 1 to 9, which is somehow difficult to effectively model the human reasoning in the decision-making processes. To deal with the vagueness and imprecision of human cognition in the MCDM problem, the most common approach is to apply fuzzy sets to model this uncertainty. Fuzzy BWM (FBWM) models the inherent ambiguity of DM in the criteria comparison process by providing the linguistic terms. These terms given by DMs then can be converted into the triangular fuzzy numbers (TFNs) as the input data of the optimization model. The detailed modelling process of FBWM can be referred to the work of Guo and Zhao (2017), Hafezalkotob and Hafezalkotob (2017), and Guo and Qi (2021) for individual and group decision scenarios.

When applying MCDM techniques to GDM scenarios, existing literature can be categorized into three main streams. Firstly, some studies focus on a single DM who represents the group or an expert team and makes the final decision (Kolagar et al., 2019). We refer to this situation as type-1 GDM. Secondly, other studies involve multiple DMs providing their individual opinions in a small decision group, which are then aggregated using specific aggregation operators (Guo & Qi, 2021). In some cases, the DM structure is hierarchical (Hafezalkotob & Hafezalkotob, 2017). We label this situation as type-2 GDM. Thirdly, in large-scale decision-making processes involving more than 20 DMs, the DMs may be divided into different groups using clustering algorithms, and group opinions are aggregated (Li & Wei, 2020). We refer to this scenario as type-3 GDM. Different GDM scenarios have the same requirement in utilizing FBWM, which is to obtain an individual Best-to-Others vector and Others-to-Worst vector.

Given the increasing attention to the impact of future events and uncertainties, scholars have emphasized the need to explore the capabilities of advanced MCDM techniques under CST (Asadabadi, 2018; Ecer & Torkayesh, 2022). It is necessary to develop a comprehensive decision support framework that considers the advanced capabilities of FBWM under CST in GDM scenarios.

3 Decision Support Framework of Stratified Fuzzy Group Best Worst Method

3.1 Definition of the Decision Problem

Determining criteria weights is a critical step in the MCDM process, as they can significantly impact the final decision. Within the context of MCDM, this paper addresses the decision problem of determining crisp weights for multiple criteria based on the subjective judgments provided by multiple DMs in two types of uncertainties: (i) DMs encounter difficulty in providing numerical preferences for the criteria through pairwise comparisons, and (ii) DMs experience uncertainty when providing their preferences due to the potential influence of future events on the relative importance of these criteria.

Based on the defined decision problem, there is a need for a structured and easily applicable group decision-making framework that can effectively synthesize individual opinions and produce optimal crisp criteria weights, while considering the inherent ambiguity of human judgments and the impact of future events.

3.2 Proposed Decision Support Framework

To address the decision problem presented earlier, this study proposes a decision support framework that focuses on subjective weight determination based on multiple DMs' pairwise comparison-based judgments under uncertain environment. The following subsection outlines the step-by-step process of the proposed decision support framework, as presented in Fig. 2.

Preliminaries

The basic definition of fuzzy sets and TFNs are presented below.

Definition 1: Given a universe of discourse X. Membership function defines a fuzzy set that maps elements to degrees of membership within a unit interval, which is [0,1]. Define $\mu_{\tilde{a}}$ is a map and $\mu_{\tilde{a}} : X \rightarrow [0, 1]$, and then each element x from X can be mapped to a crisp value in the unit interval [0,1]. $\mu_{\tilde{a}}(x)$ is the membership function of $\tilde{a}$. Fuzzy set is $\tilde{a} = \{(x, \mu_{\tilde{a}}(x)), x \in R\}$.

Definition 2: A TFN can be represented as the triple (l, m, u). It meets that $l < m < u$ and all of them are crisp value. A fuzzy number $\tilde{a}$ on R is defined to be a TFN if the membership function $\mu_{\tilde{a}}(x) : R \rightarrow [0, 1]$ is equal to Eq. (1).

$$\mu_{\tilde{a}}(x) = \begin{cases} 0,\ x < l \\ \frac{x-l}{m-l},\ l \leq x < m \\ \frac{u-x}{u-m},\ m \leq x \leq u \\ 0,\ x > u \end{cases} \quad (1)$$

Definition 3: $\tilde{a}_1$ and $\tilde{a}_2$ are any two TFNs, the algebraic operations between two TFNs are defined as in Table 1.

Table 1. Algebraic operations between two fuzzy numbers

Algebraic operator	Equation
Addition operator	$\tilde{a}_1 + \tilde{a}_2 = l_1 + l_2, m_1 + m_2, u_1 + u_2$
Subtraction operator	$\tilde{a}_1 - \tilde{a}_2 = l_1 - u_2, m_1 - m_2, u_1 - l_2$
Multiplication operator	$\tilde{a}_1 \times \tilde{a}_2 = l_1 \times l_2, m_1 \times m_2, u_1 \times u_2$
Arithmetic operator	$k \times \tilde{a}_1 = (k \times l_1, k \times m_1, k \times m_1), (k > 0)$
Defuzzification style: graded mean integration representation (GMIR)	$\mathrm{R}(\tilde{a}_i) = (l_i + 4m_i + u_i)/6$

Steps of Proposed SF-GBWM.

A key step in extending FBWM into the GDM environment is to synthesize the individual vectors through the formulas $\tilde{a}_{ij} = \tilde{a}_{Bj}/\tilde{a}_{Bi}$ and $\tilde{a}_{ij} = \tilde{a}_{iW}/\tilde{a}_{jW}$ (Guo & Qi, 2021). Additionally, input data-based consistency ratio and output data-based consistency are embedded in this framework to ensure the consistency of decision results. Each step for utilizing SF-GBWM are demonstrated as follows:

***Step 1*:** In first step, the key criteria $\{C_1, C_2, \cdots, C_n\}$ should be identified. The relevant criteria are usually determined through literature review and the outputs of domain expert.

***Step 2*:** Identify events that seem likely to occur in the future and impact the weights of the criteria system. In this step, by elucidating the impact of future events on the importance of criteria, DMs can more systematically assess the importance of these criteria in conjunction with real-world occurrences.

***Step 3*:** Assess the transition probabilities between scenarios to build the transition probability matrix based by group. In this step, the probabilities for transitioning between scenarios are assessed to build the transition probability matrix. The events with low probabilities can be ignored considering a predetermined threshold. The decision group can determine the likelihood of the occurrence of each scenario based on historical data and their expertise.

***Step 4*:** Conduct the pairwise comparisons for the best and worst criteria in each state by individual. The best criterion C_B and the worst criterion C_W from a set of criteria $\{C_1, C_2, \cdots, C_n\}$ in each state are selected in this step by individual. Then the best-compared to-others vector $\tilde{A}_B = \{\tilde{a}_{B1}, \tilde{a}_{B2}, \cdots, \tilde{a}_{Bn}\}$ and the others-compared to-worst vector $\tilde{A}_W = \{\tilde{a}_{1W}, \tilde{a}_{2W}, \cdots, \tilde{a}_{nW}\}^T$ in each state provided by individual can be obtained, in which $\tilde{a}_{Bj}$ and $\tilde{a}_{jW}$ represent the fuzzy preference of best criterion over criterion j and criterion j over the worst criterion, respectively. It should be noted that fuzzy preferences are represented as TFNs $\left(l_j, m_j, u_j\right)$.

***Step 5*:** Transform DM's pairwise comparison vectors when there is an inconsistent best criterion or worst criterion selected in GDM. The formula $\tilde{a}_{ij} = \tilde{a}_{Bj}/\tilde{a}_{Bi}$ and $\tilde{a}_{ij} = \tilde{a}_{iW}/\tilde{a}_{jW}$ can be used to convert the specific vector provided by individual.

***Step 6*:** Aggregate the comparison vectors of each DM after checking the approximate input-based consistency level according to the consistency threshold provided in the work of Guo and Qi (2021). In this step, the suitable aggregation operators such as geometric

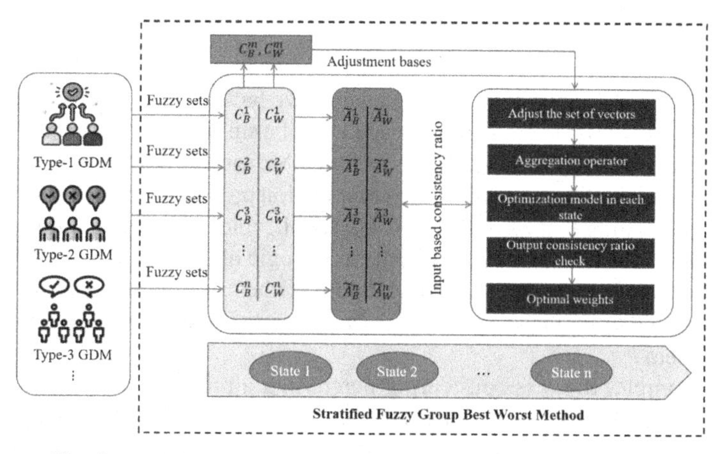

Fig. 2. Decision support framework of the proposed SF-GBWM.

mean method and the ordered weighted averaging aggregation operator can be used to integrate these individual preferences.

***Step 7*:** Calculate the weights of the criteria $\left(\tilde{w}_1^*, \tilde{w}_2^*, \cdots, \tilde{w}_n^*\right)$ in *kth* state by building a mathematical optimization model that incorporates the inherent ambiguity of expert group in the criteria comparison process. Consistent with the optimization objective of FBWM, SF-GBWM under each possible event is to obtain a single optimal solution in order to minimise the absolute difference $\tilde{\xi}$, subject to the corresponding constraints in *kth* state, namely (i)$\left|\tilde{w}_B - \tilde{a}_{Bj}\tilde{w}_j\right| \leq \tilde{\xi}$, (ii) $\left|\tilde{w}_j - \tilde{a}_{jW}\tilde{w}_W\right| \leq \tilde{\xi}$, (iii) $\sum_j R(\tilde{w}_j) = 1$ and (iv) $0 \leq l_j \leq m_j \leq u_j$, where for all $j \in [1, n]$. The steps of checking the acceptable output data-based threshold have been presented in the published works (Guo & Zhao, 2017; Liang et al., 2020; Guo & Qi, 2021).

***Step 8*:** Multiply the weights considering all possible eventualities by using the transition probability matrix. In this step, the weights of the criteria under each state obtained in last step need to be synthesized. The weights for criteria in *kth* state $S : \{S_1, \cdots, S_h\}$ are$W_k : \{w_{k1}, \cdots, w_{km}\}$. The transition matrix that includes all of the transition probabilities are shown in Eq. (2). The final optimal weights of each criterion can be computed by Eq. (3).

$$P = \begin{matrix} \\ S_1 \\ S_2 \\ \vdots \\ S_h \end{matrix} \begin{matrix} S_1 \; S_2 \; \ldots \; S_h \\ \begin{bmatrix} P_{11} & P_{12} & \ldots & P_{1h} \\ P_{21} & \vdots & & \ddots \vdots \\ \vdots & \ddots & \ddots & \vdots \\ P_{h1} & \ldots & \ldots & P_{hh} \end{bmatrix} \end{matrix} \tag{2}$$

$$W_j = \sum\nolimits_{j=1}^{h} W_k P_{kj} \tag{3}$$

4 Numerical Example

In this section, given the limited space available, we have chosen a numerical example representing type-2 GDM scenario (i.e., each DM of a small group gives their individual opinions) to provide an illustration. In this example, we assume that the importance levels of DMs are the same so as to focus on the computation process of SF-GBWM. However, it is worth mentioning that SF-GBWM is also applicable when incorporating different importance levels of DMs. Additionally, the main logic of utilizing SF-GBWM remains the same for type-1 and type-3 GDM scenarios. Other algorithms, such as clustering and weighting the importance of DMs, can be integrated with SF-GBWM based on the specific problem.

In this example, Barry is considering purchasing a property. There are three DMs involved: Barry (DM1), Barry's father (DM2), and Barry's mother (DM3). In evaluating the property, the following four criteria are taken into consideration: location (C1), condition and age (C2), size and layout (C3), and price (C4). During the determination of criteria weights, they face a high level of uncertainty due to two potential events that may impact the importance levels of the criteria. These two potential events are: Event A: A decrease in the operating condition of Barry's company, leading to unemployment. Event B: Barry getting married within the next year. The probability of neither event occurring is nearly zero, so we can exclude that state in this case. As a result, there are three possible states that encompass the separate or combined occurrence of these events. Event A occurs (S_1), Event B occurs (S_2), and both events occur simultaneously (S_3). Assume that Barry and his parents suggested probabilities for these events: a 30% likelihood for Event A, a 90% likelihood for Event B. Let the probability of S_1 be represented as P_1, and then the probabilities of S_2 and S_3 are computed as $3P_1$ and $3P_1^2$ respectively. Consequently, P_1 can be computed using the expression $4P_1 + 3P_1^2 = 1$. The computed probabilities for transitioning between states are 0.215, 0.645, and 0.139 respectively. Table 2 displays the corresponding TFNs based on the linguistic term sets. Table 3–4 presents the primary values of three DMs' opinions under three states. Based on this given dataset, the detailed computation process in state 1 is presented below. It is important to note that the computation process remains the same in each state.

Table 2. Transformation rules of linguistic variables of DMs.

Linguistic terms	Triangular fuzzy numbers
Equally importance (EI)	(1,1,1)
Weakly important (WI)	(2/3, 1, 3/2)
Fairly important (FI)	(3/2, 2, 5/2)
Very important (VI)	(5/2, 3, 7/2)
Absolutely important (AI)	(7/2, 4, 9/2)

In state 1, as the two best and worst criteria that DMs choose the most are C3 and C1, the opinion of DM 3 related to Best-to-others vector needs to be transformed using

Table 3. Best-to-others vectors: DM1-DM3 in each state.

Best Criteria			
States	S1 (0.215)	S2 (0.645)	S3 (0.139)
Others	**C3, C3, C4**	**C3, C4, C4**	**C4, C4, C4**
C1	AI, AI, AI	VI, VI, VI	AI, AI, VI
C2	FI, VI, FI	AI, AI, FI	VI, VI, FI
C3	EI, EI, VI	EI, FI, FI	FI, VI, AI
C4	FI, FI, EI	FI, EI, EI	EI, EI, EI

Notes: The bolded characters in second row represent the best criteria selected by the individual DMs. For instance, in state S1, the best criterion selected by the three DMs are C3, C3, and C4, respectively. From the third row onwards, the table displays the pairwise comparison results for each DM's best criterion over the other criteria

Table 4. Others-to-worst vectors: DM1-DM3 in each state.

Worst Criterion			
States	S1	S2	S3
Others	**C1, C1, C1**	**C2, C2, C2**	**C1, C1, C1**
C1	EI, EI, EI	WI, VI, FI	EI, EI, EI
C2	WI, WI, WI	EI, EI, EI	VI, VI, VI
C3	AI, AI, VI	AI, FI, VI	FI, FI, VI
C4	FI, VI, AI	AI, AI, AI	VI, AI, AI

Notes: The bolded characters in the second row represent the worst criteria selected by the individual DMs. For instance, in state S1, the worst criterion selected by all three DMs is C1. From the third row onwards, the table displays the pairwise comparison results for each DM's other criteria over their respective worst criterion

formula $\tilde{a}_{ij} = \tilde{a}_{Bj}/\tilde{a}_{Bi}$ and $\tilde{a}_{ij} = \tilde{a}_{iW}/\tilde{a}_{jW}$. The converted Best-to-others vector of DM 3 in state 1 are shown below:

$$\left\{\begin{array}{c} \tilde{A}^3_{Btr} = \left(\tilde{a}^3_{Btr1}, \tilde{a}^3_{Btr2}, \tilde{a}^3_{Btr3}, \tilde{a}^3_{Btr4}\right) \\ = \left(\tilde{a}^3_{31}, \tilde{a}^3_{32}, \tilde{a}^3_{33}, \tilde{a}^3_{34}\right) \\ = \left(\frac{\tilde{a}^3_{41}}{\tilde{a}^3_{43}}, \frac{\tilde{a}^3_{42}}{\tilde{a}^3_{43}}, (1, 1, 1), \frac{\tilde{a}^3_{44}}{\tilde{a}^3_{43}}\right) \\ = ((1, 1.33, 1.8), (0.43, 0.67, 1), (1, 1, 1), (0.29, 0.33, 0.4)) \end{array}\right\} \quad (4)$$

Before integrating the comparison vectors of DMs, it is necessary to check the input data-based consistency ratio for each vector provided by each DM. The detailed computation process for the input data-based consistency ratio has been presented in the published work (Guo & Qi, 2021). If the input data-based consistency ratio of a certain DM is not acceptable, the DM will be advised to adjust their opinion accordingly. Once each DM's input satisfies the consistency requirement, the suitable aggregation operator can be used to integrate the individual opinions of the DMs based on the specific scenario. In this example, individual input data-based consistency ratio in state 1 is below the threshold, which means the input data is of high consistency level and can be used for further computation. Among various aggregation operators in GDM, the geometric mean method is one of the simplest and efficient method to average all the judgements provided by DMs (Guo & Qi, 2021). Technically speaking, assuming that there are q DMs, the final aggregated vector $\tilde{a}_{Bj}$ for group preference vector can be expressed as $\tilde{a}_{Bj} = \left(\tilde{a}^1_{Bj} \otimes \tilde{a}^2_{Bj} \otimes \cdots \otimes \tilde{a}^q_{Bj}\right)^{1/q}$. Since the geometric mean operation is less affected by extreme value in GDM scenarios and easy for use, this operator is utilized here to synthesize the individual preferences. The final two integrated vectors in state 1 can be obtained as follows.

$$\tilde{A}_{B^{tr}} = \left(\tilde{a}_{B^{tr}1}, \tilde{a}_{B^{tr}2}, \tilde{a}_{B^{tr}3}, \tilde{a}_{B^{tr}4}\right)$$

$$= ((2.31, 2.77, 3.32), (1.17, 1.59, 2.06), (1, 1, 1), (0.87, 1.10, 1.36)) \tag{5}$$

$$\tilde{A}_{W^{tr}} = \left(\tilde{a}_{1W^{tr}}, \tilde{a}_{2W^{tr}}, \tilde{a}_{3W^{tr}}, \tilde{a}_{4W^{tr}}\right)$$

$$= ((1, 1, 1), (0.67, 1, 1.5), (3.13, 3.63, 4.14), (2.36, 2.88, 3.40)) \tag{6}$$

Based on the input data, the FBWM optimization model shown in Step 7 (Guo & Zhao, 2017; Guo & Qi, 2021) is constructed using Lingo Version 18.0 software in each state. The output data-based consistency ratio should be checked in each computation process. By following the same transformation and consistency check logic, the optimal weights of each criterion in each state can be obtained, as shown in Table 5. Subsequently, using the GMIR in Table 1, the TFNs in Table 5 can be transformed into crisp values. For example, the crisp value of the optimal weight of C1 in S1 is (0.1011 + 0.1164 + 0.134*1)/6 $\approx$ 0.117. Following the same computation procedure, the crisp value of each criterion in each state can be calculated. Still taking C1 as an example, the optimal weight of C1 is computed by considering the probability of event occurrence (i.e., the transition probabilities of S1, S2, and S3 are 0.215, 0.645, and 0.139, respectively). The final optimal weight of C1 is obtained as 0.117*0.215 + 0.190*0.645 + 0.107*0.139 $\approx$0.163. Similarly, the optimal weights of C2, C3, and C4 can be calculated. The optimal weights of the evaluation criteria are presented in Table 6, which also indicates that, considering the two potential future events in this example, the criteria are prioritized in the following order: price (C4) $\succ$ size and layout (C3) $\succ$ location (C1) $\succ$ condition and age (C2).

Through this straightforward computation process, individual opinions are synthesized into group opinions while considering the ambiguity of human decision-making and the uncertainty influenced by possible future events. This leads to a more rational decision-making process. However, one point that can be criticized is the repeated computation process as the number of DMs increases and the possible states stemming from future events. To address this, developing an automated process for the above computations could be essential for the real-life implementation.

Table 5. Fuzzy optimal weights of criteria using SF-BWGM in each state

Criteria	C1	C2	C3	C4	ξ
Weight in S1	(0.101, 0.116, 0.134)	(0.189, 0.189, 0.189)	(0.320, 0.320, 0.320)	(0.308, 0.381, 0.416)	0.022
Weight in S2	(0.146, 0.188, 0.243)	(0.100, 0.112, 0.128)	(0.278, 0.355, 0.401)	(0.348, 0.348, 0.348)	0.022
Weight in S3	(0.069, 0.100, 0.175)	(0.150, 0.208, 0.346)	(0.137, 0.189, 0.309)	(0.376, 0.453, 0.639)	0.020

Table 6. Crisp optimal weights of criteria using SF-BWGM in each state and the final output optimal weights

Criteria	C1	C2	C3	C4
Weight in S1	0.117	0.189	0.320	0.375
Weight in S2	0.190	0.113	0.350	0.348
Weight in S3	0.107	0.221	0.200	0.471
Optimal Weight	**0.163**	**0.144**	**0.322**	**0.371**

5 Conclusion

In this study, the proposed SF-GBWM is specifically designed and well-suited for determining the optimal criteria weights in GDM scenarios that consider future events and the uncertainty arising from human subjective judgments, while reducing the number of pairwise comparison steps through a structured best-worst-based comparison process. To address the inherent ambiguity of human-involved decision-making, the pairwise comparison-based preferences provided by DMs are articulated via linguistic terms. Fuzzy sets are employed to convert DMs' linguistic preferences into computable fuzzy numbers. To account for the uncertainty associated with the potential state of a decision, the main idea of CST is utilized to elucidate the stratified decision-making process. The current state of a decision is identified, and potential states that may occur and are adjacent to the current state are also factored into the decision-making process. The proposed approach can be applied in various domains, such as business, engineering, or healthcare, where criteria weights are required to be determined collaboratively by a group of DMs or experts under uncertain environments.

One limitation of this research lies in the fact that the proposed method is based on type-1 fuzzy sets. Future studies may consider incorporating other types of fuzzy sets. Furthermore, conducting more comparative analyses of different MCDM techniques under CST within a fuzzy GDM environment would be beneficial. Additionally, future studies could develop application software to assist the decision-making process. Furthermore, there is potential for further practical applications to be explored.

Acknowledgments. The authors would like to thank the Research and Innovation Office of the Hong Kong Polytechnic University for supporting the project (Project Code: RKQY).

References

Asadabadi, M.R.: The stratified multi-criteria decision-making method. Knowl.-Based Syst. **162**, 115–123 (2018). https://doi.org/10.1016/j.knosys.2018.07.002

Asadabadi, M.R., Zwikael, O.: Integrating risk into estimations of project activities' time and cost: a stratified approach. Eur. J. Oper. Res. **291**(2), 482–490 (2021). https://doi.org/10.1016/j.ejor.2019.11.018

Asadabadi, M.R., Ahmadi, H.B., Gupta, H., Liou, J.J.: Supplier selection to support environmental sustainability: the stratified BWM topsis method. Ann. Oper. Res. **322**(1), 321–344 (2023). https://doi.org/10.1007/s10479-022-04878-y

Ecer, F., Torkayesh, A.E.: A stratified fuzzy decision-making approach for sustainable circular supplier selection. IEEE Trans. Eng. Manage. **71**, 1130–1144 (2022). https://doi.org/10.1109/tem.2022.3151491

Guo, S., Zhao, H.: Fuzzy best-worst multi-criteria decision-making method and its applications. Knowl.-Based Syst. **121**, 23–31 (2017). https://doi.org/10.1016/j.knosys.2017.01.010

Guo, S., Qi, Z.: A fuzzy best-worst multi-criteria group decision-making method. IEEE Access **9**, 118941–118952 (2021). https://doi.org/10.1109/access.2021.3106296

Hafezalkotob, A., Hafezalkotob, A.: A novel approach for combination of individual and group decisions based on fuzzy best-worst method. Appl. Soft Comput. **59**, 316–325 (2017). https://doi.org/10.1016/j.asoc.2017.05.036

Kolagar, M., Hosseini, S.M., Felegari, R., Fattahi, P.: Policy-making for renewable energy sources in search of sustainable development: a hybrid DEA-FBWM approach. Environ Syst. Decisions **40**(4), 485–509 (2019). https://doi.org/10.1007/s10669-019-09747-x

Li, S., Wei, C.: A large scale group decision making approach in healthcare service based on sub-group weighting model and hesitant fuzzy linguistic information. Comput. Ind. Eng. **144**, 106444 (2020). https://doi.org/10.1016/j.cie.2020.106444

Rezaei, J.: Best-worst multi-criteria decision-making method. Omega **53**, 49–57 (2015). https://doi.org/10.1016/j.omega.2014.11.009

Torkayesh, A.E., Malmir, B., Rajabi Asadabadi, M.: Sustainable waste disposal technology selection: The stratified best-worst multi-criteria decision-making method. Waste Manage. **122**, 100–112 (2021). https://doi.org/10.1016/j.wasman.2020.12.040

Zadeh, L.A.: Fuzzy sets. Inf. Control. **8**(3), 338–353 (1965). https://doi.org/10.1016/s0019-9958(65)90241-x

Zadeh, L.A.: Stratification, target set reachability and incremental enlargement principle. Inf. Sci. **354**, 131–139 (2016). https://doi.org/10.1016/j.ins.2016.02.047

Selection of Rapid Classifier Development Methodology Used to Implement a Screening Study Based on Children's Behavior During School Lessons

Grzegorz Dziczkowski[1,2], Tomasz Jach[1(✉)], Barbara Probierz[1,2], Piotr Stefanski[1], and Jan Kozak[1,2]

[1] Department of Machine Learning, University of Economics in Katowice, 1 Maja 50,40-287 Katowice, Poland
{jan.kozak,jan.kozak}@ue.katowice.pl
[2] Łukasiewicz Research Network–Institute of Innovative Technologies EMAG, Leopolda 31, 40-189 Katowice, Poland

Abstract. The purpose of the article is to prepare a methodology for an advanced system that implements screening among children of early school age. The screening will be implemented in classrooms using cameras. Cameras in bi-weekly windows will study children's behavior and the system will report alerts when abnormal behavior is detected. The alerts are intended to recommend in-depth examinations with a specialist. In this article, the authors present a preliminary study to assess the feasibility of rapidly creating classifiers that detect specific behavioral elements (e.g., open mouth, putting fingers in mouth, asymmetrical closing of eyes, etc.). The article aims to define a methodology for detecting anomalies in children's behavior, which in the next stages of the project will be used to detect undesirable behaviors such as lack of concentration, hyperactivity, epilepsy, undesirable behavior to noise and stress. The aim of the presented research is to create a methodology based on proprietary neural network-based classifiers in later studies implementing screening tests. The presented article presents research comparing the performance of two different neural network architectures: an advanced ResNet-based model and a simpler custom convolutional neural network (CNN). The research presented here demonstrates that both advanced and simple models have their place in the rapid development of microclassifiers and allow acceptance of the chosen methodology for further work on student screening.

Keywords: Computer Vision · Behavior analysis · machine learning · knowledge discovery in databases

1 Introduction

The article aims to validate the proposed methodology realizing screening based on children's behavior during lessons. The presented methodology is aimed at rapid selection of classifiers realizing selection of children's behavior and identifying abnormal behavior. The target system is a research project whose goal

M. Campos Ferreira et al. (Eds.): GDN 2024, LNBIP 509, pp. 77–88, 2024.
https://doi.org/10.1007/978-3-031-59373-4_7

is to implement an innovative information system based on the latest research in the field of image analysis to realize the initial screening. The system will be installed in the classrooms of school establishments and will carry out screening for children in grades 1–3. Implementation of the screening tests will be done with a set of cameras that record children's behaviour during classroom activities. The system will analyse the psychophysical behaviour of the students. The system does not implement student identification and does not hold sensitive data. The system focusses on analysing individual body parts and analyses student behaviour based on this. During the study in the classroom, the system identifies the student by the seat occupied, so the system works under the assumption that students occupy the same seats in the classroom during the study.

The system will analyse the behaviour of all students in relevant time windows. For each time slot, the system will determine typical behaviours suitable for the current teaching process. Having patterns of behaviour suitable for different intensity of lessons conducted, the system will be able to analyse behaviours that deviate significantly from the selected pattern. Each student will be analysed for longer periods of time. In case of repeated deviations from the norm, the system will report an alert. Subsequently, the information justifying the alert filing, along with the reasons for the decision made by the system, will be forwarded to the student's parents or legal guardians to suggest consultations with psychological and pedagogical counseling centres or others, including adequate medical ones to expand further diagnosis.

The system will be able to analyse students' behavioural anomalies, which could indicate serious diseases. The timing of disease detection is very important, as the prognosis is often much more successful if treatment begins at an early stage of the disease and at a young age of the patient. This reduces the cost of treatment and increases the chances of recovery. The system will focus on the detection of behaviors that may indicate the following disorders: - Obsessive-compulsive disorder, e.g., repetitively checking the presence of things, taking things out and putting them in the pencil case, counting them, putting them back, - Autism spectrum disorders e.g. covering up, plugging ears, nodding, rubbing, rapid hand movements, - Anxiety/depression disorders e.g. compulsive repetitive behavior- persistent cleaning of an object, clothing, scratching oneself, nail biting, fidgeting - Epilepsy e.g. convulsions involving part of the face, - Attention deficit hyperactivity disorder e.g. fidgeting, changing positions more often in relation to peers. A suitable research methodology must be tested and selected. The authors in this article focus on preliminary work to develop a methodology for quickly creating classifiers that detect particular behaviours. In the masterworks that follow, the authors will have a large number of classifiers for detecting undesirable individual behaviours. The work presented here is aimed at presenting a methodology for creating and evaluating suitable classifiers and is a preliminary study of the presented project.

2 Theoretical Background

In this section, we provide a theoretical background for the concepts and techniques used in our research. We discuss neural networks, specifically Convolutional Neural Networks (CNNs) and Recurrent Neural Networks (RNNs), existing classification methods and their limitations, and introduce the concept of transfer learning and its applications.

2.1 Neural Networks

Neural networks are a class of machine learning algorithms inspired by the structure and functioning of the human brain. They consist of interconnected nodes, also known as neurones, organised in layers. Two types of neural networks are particularly relevant to our research:

Convolutional Neural Networks (CNNs). CNNs are a type of neural network designed to process grid-like data, such as images and videos. They are made up of multiple layers, including convolutional layers, pooling layers, and fully connected layers. CNNs have proven to be highly effective in image classification tasks because of their ability to automatically extract relevant features from images.

Recurrent Neural Networks (RNNs). RNNs are a type of neural network designed for sequential data, such as time series and text in natural language. They include recurrent connections that allow them to maintain a form of memory, making them suitable for tasks that involve sequences. RNNs have been applied in various natural language processing tasks, including sentiment analysis and language generation.

2.2 Existing Classification Methods and Limitations

Traditional classification methods include decision trees, support vector machines, and k-nearest neighbours. Although these methods have been widely used, they often have limitations in handling complex, high-dimensional data, such as images and text. CNNs and RNNs have emerged as powerful alternatives due to their ability to automatically learn hierarchical features from data.

2.3 Transfer Learning and Its Applications

Transfer learning is a machine learning technique in which a model trained on one task is adapted to a related but different task. It has gained popularity in various domains, including computer vision and natural language processing. In our research, we explore transfer learning using pre-trained models, such as ResNet, to leverage knowledge learnt from large datasets and apply it to our specific microclassification task.

3 Related Works

Various methodologies have emerged in the dynamic field of video object detection, ranging from traditional machine learning [6,27] to advanced deep learning techniques [8,19]. Human behaviour recognition technology [11] has a wide range of applications in security surveillance [4], medical diagnosis and monitoring [22], and human-computer interaction [25]. Much of current research highlights the importance of feature fusion methods in improving face recognition and detection systems [23].

Integrating CNNs (convolutional neural networks) with traditional approaches not only improves accuracy, but also expands the applicability of these systems to various datasets and scenarios, representing a significant step toward more robust and versatile facial analysis technologies [14]. Feature fusion methods, using both convolutional neural networks (CNN) and traditional techniques, have emerged as key elements in improving face detection and recognition systems [26]. In [2], a deep heterogeneous feature fusion approach is proposed using deep CNNs is proposed. This method effectively integrates features from various deep networks, leading to a significant improvement in face recognition accuracy. Subsequent research [21] presents a multimodal facial recognition system that combines CNN features with hand-crafted counterparts, achieving improved system performance according to experiments with the RGB-D-T database. The approach in [20] proposes feature fusion based on the Combination of Shifted Filter Responses (COSFIRE) and pre-trained VGG CNNs for gender recognition. Research has also been carried out on 3D facial recognition, combining 2D and 3D features [7], contributing to improving the performance of face recognition, as well as building algorithms for 3D facial reconstruction from photos and videos [12]. Subsequent research in [13] introduces covariance matrix (CMR) regularisation to mitigate overadaptation during feature fusion. The proposal in [24] uses Feature Fusion and Segmentation Supervised (DF2S2) detection for face detection, achieving state-of-the-art performance on the WIDER FACE benchmark. Other research in [15] proposes an age-independent facial recognition system by integrating the VGG model and using multidiscriminant feature-level correlation analysis.

Significant progress was made in the field of classroom face detection and attention research, where three convolutional neural networks: Det-A, Det-B and Det-C were developed [10]. These networks were inspired by the Adaboost cascading algorithm, achieving an impressive recall efficiency of 92.9%. Additionally, a prototype convolutional neural network called HeadNet incorporated facial location information across multiple frames for class size statistics. In [5] used Faster R-CNN (region-based convolutional neural network) based on a pre-trained ZFNet model to extract characteristics of students' classroom behaviour, such as studying, sleeping, and playing with mobile phones. Meanwhile, the use of the YOLO [16] deep learning model allows extraction of features for key facial points to assess student participation in the classroom [3]. Furthermore, the use of the YOLO algorithm allows the recognition of seven behaviours in the classroom, such as reading, sleeping, raising hands, writing, listening, standing, and looking left and right [9]. Research [18] introduced a face detection method that combined multiple deep neural networks to improve detection accuracy. The FaceNet face recognition method is extended with SVM to extract facial features. In the following years, the MedianFlow-based face tracking algorithm was improved using MTCNN, and a novel CNN-based facial keypoint detection model was proposed [1]. This model integrated information about eye closure time, blink rate, and head position to detect facial fatigue [17].

4 Methodology

The authors of this article present preliminary research aimed at evaluating the feasibility of rapidly developing classifiers to detect specific behavioral elements, including but not limited to open/closed mouth, finger sucking, insertion of objects into the mouth, among others. The identification of mouth breathing through pattern analysis in children serves to monitor abnormalities in their preferred respiratory pathway. Under physiological circumstances, breathing pathway occurs predominantly through the nasal cavity. However, heightened instances of mouth breathing in childhood may lead to abnormalities in subsequent growth. A pathological distribution of muscle tension during breathing can result in various consequences, including alterations in craniofacial structure, compromised lip sealing, hypotonic lips, drooping eyelids, and malocclusions like an anterior open bite. Early detection of respiratory pathway abnormalities is crucial for initiating timely and effective interventions, thereby mitigating potential associated costs.

4.1 Dataset Description

The dataset, meticulously crafted by the authors, consists of 8848 images generated from two 30-second high-definition video clips, personally recorded to capture the nuanced differences in facial state, focussing particularly on the mouth region. These videos, one depicting a subject with an open mouth and the other with a closed mouth, were recorded in a controlled environment to ensure consistency in lighting and positioning. The choice of FullHD resolution for recording ensured that the images extracted were of high quality, capturing the necessary details required for accurate classification.

During preprocessing, a critical step in standardising the dataset for the neural network, each video was processed frame by frame using the OpenCV library. This process transformed the dynamic video data into a series of static images, each of which potentially serves as an independent data point for training the classifier. The extracted frames were resized to a uniform dimension of 64x64 pixels, strategically reducing the computational burden during training while retaining the essential features necessary for classification. The processed images were then methodically organised into directories corresponding to their respective classes, 'open' and 'closed', facilitating a streamlined and efficient training process.

In addition to the standard preprocessing steps, data augmentation was employed to enhance the model's generalisability and robustness. This was achieved using a sequential Keras model that applied random horizontal flips, rotations of up to 0.1 rad, and zooms of up to 10%. These transformations introduced a healthy variation in the dataset, simulating real-world deviations and nuances, thus preparing the model for a wide range of scenarios beyond the training data.

The data set was divided into training and validation sets, with 7079 images (approximately 80%) used to train the model and the remaining 1769 images (approximately 20%) used for validation. This split ensured a comprehensive evaluation, allowing the model to learn and adapt to the patterns within the data while providing a reliable basis for assessing its performance and generalisation capabilities.

This dataset, with its careful preparation and organisation, is critical for the study's goals. The quality, quantity and meticulous preparation of the data ensure that the resulting model is not only accurate in its current form, but also robust and adaptable to similar tasks in real-world applications. The subsequent sections will delve into the algorithm and model architecture, building upon this solid foundation to achieve the study's objectives.

4.2 Model Architectures

In our study, we used neural networks, which are effective and widely used in the problems of image classification. In addition, we compare the performance of two distinct neural network architectures: ResNet50 and a simple custom-built CNN. The following are the details of each model's architecture.

ResNet50 Architecture. ResNet, short for Residual Networks, is a classic network known for its depth and ability to train hundreds or even thousands of layers successfully. The key innovation in ResNet is the introduction of the residual block. Typical ResNet models are implemented with 50, 101, or 152 layers. In our study, we used the 50-layer variant, commonly known as ResNet50.

ResNet50 consists of an initial convolutional and max-pooling layer followed by 16 residual blocks. Each residual block has a stack of 3 layers. The network ends with average pooling, a fully connected layer, and a final softmax layer. The distinguishing feature of ResNet is the skip connections or shortcuts that allow the gradient to bypass one or more layers.

Custom Simple CNN Architecture. Our custom CNN is designed to be significantly simpler than ResNet50 to assess whether such a model can perform adequately for our task. The architecture is as follows:

1. **Data Augmentation Layer:** A series of data augmentation techniques such as random flips and rotations are applied to the input data to make the model more robust to variations in the input data.

2. **Rescaling Layer:** This layer normalises the image pixels to between 0 and 1 by scaling with a factor of (1./255).
3. **Convolutional and Max-Pooling Layers:** Three sets of convolutional layers followed by max-pooling layers are used. The first convolutional layer has 16 filters, the second 32, and the third 64, each with a kernel size of 3x3 and 'same' padding. ReLU activation is used for these layers. Each convolutional layer is followed by a max-pooling layer, which reduces the spatial dimensions by half.
4. **Dropout Layer:** A dropout layer with a dropout rate of 0.2 is used after the last max-pooling layer to prevent overfitting.
5. **Flatten Layer:** This layer flattens the output of the convolutional layers to form a single long feature vector.
6. **Dense Layers:** After flattening the output, it is followed by a fully connected dense layer with 128 units and ReLU activation. The final layer is a dense layer with a single neurone and a sigmoid activation function to output the probability of a positive class.

This model is considerably less complex than ResNet50, with fewer layers and parameters. We aim to assess its performance in comparison to the more sophisticated ResNet50 to understand the trade-offs involved in choosing a more straightforward architecture for our specific task.

4.3 Algorithm Details

The core of our methodology is centred around the comparison of two distinct models: a sophisticated ResNet-based model and a simpler custom convolutional neural network (CNN). This comparative approach is designed to evaluate the effectiveness of complex versus straightforward architectures in handling our specific image classification task.

1. **Video Frame Extraction:** The algorithm starts by processing the raw video clips, extracting individual frames to be used as separate data points. This conversion from temporal video data to a series of static images is crucial as it captures specific moments in time, each representative of the mouth's state.
2. **Image Preprocessing:** Each extracted frame is resized to a uniform dimension of 64x64 pixels to ensure consistency and manageability for network input. This standardisation is essential for computational efficiency and helps to focus the model on the most relevant features within the image.
3. **Data Augmentation:** To enhance the robustness of both models, data augmentation techniques such as random flips, rotations, and zooms are applied. These techniques introduce variability in the dataset, preparing the models for a wider range of real-world scenarios.
4. **Model Training:** The preprocessed and augmented images are fed into two models for training and evaluation:
   - **ResNet Model:** The first model is based on ResNet50, known for its deep architecture and ability to learn complex patterns. ResNet is particularly chosen for its residual learning framework, which addresses the

degradation problem and enables the training of substantially deeper networks. This depth is beneficial to capture the nuanced differences between open and closed mouths.
- **Simple CNN Model:** The second model is a simpler custom CNN composed of convolutional, max-pooling, and dense layers. This model is included to assess whether a less complex architecture can adequately perform the task, providing a comparison with the more sophisticated ResNet50.

The decision to utilise the ResNet algorithm alongside a simpler CNN model was driven by the desire to understand the trade-offs between complexity and performance in our specific task. While ResNet's deep architecture and residual learning capabilities make it a powerful tool for image classification, it is also important to evaluate whether simpler models can achieve comparable results with potentially lower computational costs and easier interpretability. The comparative analysis of these two models will provide insights into the most effective and efficient approaches for classifying the mouth's state from video frames.

4.4 Training and Validation Process

The training and validation are crucial steps in ensuring the precision and generalisability of every model. We employed a systematic approach to train both the ResNet and the simple custom CNN models using our prepared data set. The process is detailed as follows.

Early Stopping: To prevent overfitting and optimise the training time for both models, we implemented an early stopping mechanism. This technique monitors the loss on the training dataset and stops the training process if the loss doesn't improve after three consecutive epochs. Specifically, we used the EarlyStopping callback from TensorFlow, configured to monitor the 'loss' metric with a patience of 3.

Training the Models: Both models were trained for a maximum of 10 epochs, a period we found sufficient for them to learn from the data without overfitting. During training, the weights of each model were updated using the backpropagation algorithm, with the training dataset fed in batches. The performance of each model was continuously monitored using the validation set.

Performance Metrics: To evaluate each model's performance during and after training, we tracked several key metrics:

- **Accuracy:** Both training and validation accuracy were monitored for each model to understand how well they were learning and generalizing.
- **Loss:** Training and validation loss for each model were tracked to understand how each model's predictions deviated from the actual labels over time.

Visualization: To visually interpret each model's performance, we plotted the training and validation accuracy and loss over the epochs for both the ResNet model and the simple custom CNN. This graphical representation provides an intuitive understanding of the learning process, helping identify when each model has learnt optimally or if it is beginning to overfit. The comparative results are discussed in the following section.

Model Evaluation: Finally, we evaluated the performance of both models in the validation set using a classification report. This report provided detailed metrics such as precision, recall, and F1 score for each class, giving a comprehensive view of each model's capabilities. Predictions were made in the validation set and predicted labels were compared against true labels to calculate these metrics for both models.

This structured approach ensures that both models are not only accurate but also robust and reliable and capable of performing well in real-world applications beyond the validation set. Comparative analysis between the complex ResNet and the simpler CNN will provide valuable insight into the trade-offs involved in model selection for this specific task.

5 Conclusions and Results

We sought to compare the efficiency and ease of developing microclassifiers using a sophisticated ResNet50 model and a simpler custom-built Convolutional Neural Network (CNN). Here, we present the results and conclusions that highlight the speed and ease of development for each model.

5.1 ResNet50 Results

The ResNet50 model showcased not only high performance but also the potential for rapid development of microclassifiers. Despite its deep and complex architecture, the model quickly adapted to the dataset, showing significant improvement in early epochs. This rapid learning curve demonstrates ResNet50's capability for fast deployment in microclassifier creation, achieving a final training accuracy of approximately 97.73% and perfect accuracy on the validation set within just 7 epochs.

5.2 Custom CNN Results

The custom CNN represents the epitome of ease in creating microclassifiers. With a simpler structure, it was quick to implement and train. Although it did not match ResNet50's performance, it achieved a commendable final training accuracy of approximately 89.55% and validation accuracy of approximately 91. 63%. Its less complex nature allows for faster iterations and modifications, making it an ideal candidate for scenarios where speed and ease of development are crucial.

5.3 Comparative Analysis

Table 1. Comparative Results of ResNet50 and Custom CNN

Metric	ResNet50	Custom CNN
Final Training Accuracy	0.98	0.90
Final Validation Accuracy	1.00	0.92
Test Accuracy	1.00	0.92
Test Precision (Open)	1.00	0.91
Test Precision (Closed)	1.00	0.92
Test Recall (Open)	1.00	0.91
Test Recall (Closed)	1.00	0.92
Test F1-Score (Open)	1.00	0.91
Test F1-Score (Closed)	1.00	0.92

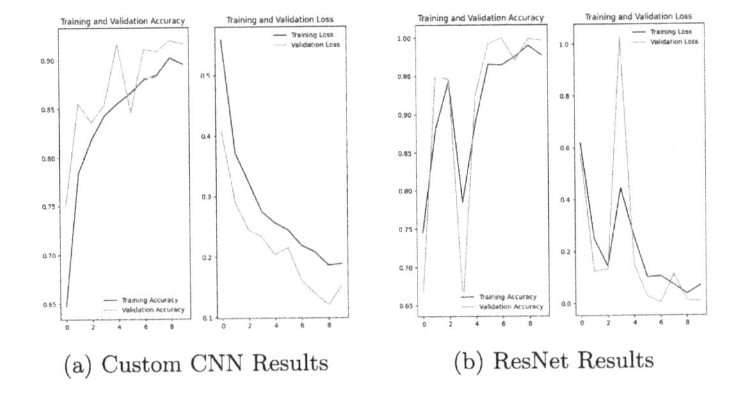

(a) Custom CNN Results (b) ResNet Results

Fig. 1. Comparison of Results

The comparative analysis between ResNet50 and custom CNN highlights a trade-off between complexity and speed. ResNet50, while more complex, provides a rapid and powerful solution for microclassifier development. On the contrary, the custom CNN, with its simpler architecture, allows for quicker development cycles and easier modifications, which is beneficial for rapid prototyping and scenarios with limited computational resources (Fig. 1 and Table 1).

The "Training and Validation Accuracy" and "Training and Validation Loss" plots further support this analysis. They show that while ResNet50 achieves higher performance more quickly, the custom CNN still maintains a steady and relatively rapid improvement, underlining its suitability for fast-paced development environments.

6 Conclusion

The timing of disease detection is very important, as the prognosis is often much more successful if treatment begins at an early stage of the disease and at a young age of the patient. This reduces the cost of treatment and increases the chances of recovery. The results of the work proved that the proposed method is able to detect an open mouth in a monitored person, which is the first step in the process of detecting many respiratory diseases.

In addition, this study has demonstrated that both sophisticated and simple models have their place in the rapid creation of microclassifiers. ResNet50 stands out for its rapid adaptation and high performance, making it suitable for applications where accuracy is paramount and some complexity is acceptable. On the other hand, custom CNN, with its straightforward architecture and decent performance, exemplifies ease and speed in development, ideal for situations requiring rapid deployment and iteration.

Future research could explore combining the strengths of both models, potentially creating a hybrid system that balances the ease of development with high performance, furthering the field of rapid microclassifier creation.

References

1. Bi, X., Chen, Z., Yue, J.: A Novel one-step method based on YOLOv3-tiny for fatigue driving detection. In: 2020 IEEE 6th International Conference on Computer and Communications (ICCC), pp. 1241–1245. IEEE (2020)
2. Bodla, N., Zheng, J., Xu, H., Chen, J.C., Castillo, C., Chellappa, R.: Deep heterogeneous feature fusion for template-based face recognition. In: 2017 IEEE Winter Conference on Applications of Computer Vision (WACV), pp. 586–595. IEEE (2017)
3. Chen, H., Zhou, G., Jiang, H.: Student behavior detection in the classroom based on improved YOLOv8. Sensors **23**(20), 8385 (2023)
4. Cheng, Y.: Video-based student classroom classroom behavior state analysis. Int. J. Educ. Humanit. **5**(2), 229–233 (2022)
5. Cowton, J., Kyriazakis, I., Bacardit, J.: Automated individual pig localisation, tracking and behaviour metric extraction using deep learning. IEEE Access **7**, 108049–108060 (2019)
6. Dhillon, A., Verma, G.K.: Convolutional neural network: a review of models, methodologies and applications to object detection. Prog. Artif. Intell. **9**(2), 85–112 (2020)
7. Feng, J., Guo, Q., Guan, Y., Wu, M., Zhang, X., Ti, C.: 3D face recognition method based on deep convolutional neural network. In: Panigrahi, B.K., Trivedi, M.C., Mishra, K.K., Tiwari, S., Singh, P.K. (eds.) Smart Innovations in Communication and Computational Sciences. AISC, vol. 670, pp. 123–130. Springer, Singapore (2019). https://doi.org/10.1007/978-981-10-8971-8_12
8. Han, Y.K., Choi, Y.B.: Human action recognition based on LSTM model using smartphone sensor. In: 2019 Eleventh International Conference on Ubiquitous and Future Networks (ICUFN), pp. 748–750. IEEE (2019)

9. He, K., Zhang, X., Ren, S., Sun, J.: Deep residual learning for image recognition. In: Proceedings of the IEEE Conference on Computer Vision and Pattern Recognition, pp. 770–778 (2016)
10. Jiang, B., Xu, W., Guo, C., Liu, W., Cheng, W.: A classroom concentration model based on computer vision. In: Proceedings of the ACM Turing Celebration Conference-China, pp. 1–6 (2019)
11. Khan, M.A., et al.: Human action recognition using fusion of multiview and deep features: an application to video surveillance. Multimedia Tools and Appl. **83**(5), 1–27 (2020)
12. La Cava, S.M., Orrù, G., Drahansky, M., Marcialis, G.L., Roli, F.: 3D face reconstruction: the road to forensics. ACM Comput. Surv. **56**(3), 1–38 (2023)
13. Lu, Z., Jiang, X., Kot, A.: Feature fusion with covariance matrix regularization in face recognition. Signal Process. **144**, 296–305 (2018)
14. Mandal, B., Okeukwu, A., Theis, Y.: Masked face recognition using resnet-50. arXiv preprint arXiv:2104.08997 (2021)
15. Moustafa, A.A., Elnakib, A., Areed, N.F.: Age-invariant face recognition based on deep features analysis. SIViP **14**, 1027–1034 (2020)
16. Redmon, J., Divvala, S., Girshick, R., Farhadi, A.: You only look once: unified, real-time object detection. In: Proceedings of the IEEE Conference on Computer Vision and Pattern Recognition, pp. 779–788 (2016)
17. Ren, H., et al.: A real-time and long-term face tracking method using convolutional neural network and optical flow in IoT-based multimedia communication systems. Wirel. Commun. Mob. Comput. **2021**, 1–15 (2021)
18. Schroff, F., Kalenichenko, D., Philbin, J.: FaceNet: A unified embedding for face recognition and clustering. In: Proceedings of the IEEE Conference on Computer Vision and Pattern Recognition, pp. 815–823 (2015)
19. Sharma, V., Gupta, M., Kumar, A., Mishra, D.: Video processing using deep learning techniques: a systematic literature review. IEEE Access **9**, 139489–139507 (2021)
20. Simanjuntak, F., Azzopardi, G.: Fusion of CNN- and COSFIRE-based features with application to gender recognition from face images. In: Arai, K., Kapoor, S. (eds.) CVC 2019. AISC, vol. 943, pp. 444–458. Springer, Cham (2020). https://doi.org/10.1007/978-3-030-17795-9_33
21. Simón, M.O., et al.: Improved RGB-D-T based face recognition. Iet Biometrics **5**(4), 297–303 (2016)
22. Sun, Z., Ke, Q., Rahmani, H., Bennamoun, M., Wang, G., Liu, J.: Human action recognition from various data modalities: a review. IEEE Trans. Pattern Anal. Mach. Intell. **45**(3), 3200–3225 (2022)
23. Taskiran, M., Kahraman, N., Erdem, C.E.: Face recognition: past, present and future (a review). Digital Signal Proc. **106**, 102809 (2020)
24. Tian, W., et al.: Learning better features for face detection with feature fusion and segmentation supervision. arXiv preprint arXiv:1811.08557 (2018)
25. Verma, Kamal Kant, Singh, Brij Mohan, Dixit, Amit: A review of supervised and unsupervised machine learning techniques for suspicious behavior recognition in intelligent surveillance system. Int. J. Inf. Technol. 1–14 (2019). https://doi.org/10.1007/s41870-019-00364-0
26. Zhao, Z., Alzubaidi, L., Zhang, J., Duan, Y., Gu, Y.: A comparison review of transfer learning and self-supervised learning: definitions, applications, advantages and limitations. Expert Syst. Appl. 122807 (2023)
27. Zhu, H., Wei, H., Li, B., Yuan, X., Kehtarnavaz, N.: A review of video object detection: datasets, metrics and methods. Appl. Sci. **10**(21), 7834 (2020)

Upper Performance Limits and Distribution Invariance for Surrogate Weights in MCDA

Sebastian Lakmayer[1], Mats Danielson[1,2](✉), and Love Ekenberg[2,1]

[1] Department of Computer and Systems Sciences, Stockholm University, PO Box 7003, SE-164 07 Kista, Sweden
mats.danielson@su.se

[2] International Institute for Applied Systems Analysis, IIASA, Schlossplatz 1, AT-2361 Laxenburg, Austria

Abstract. This paper proposes a simple approach to determining the upper performance limit (approximate the maximum hit ratio) for surrogate weight models in additive models of multi-criteria decision analysis. The approach is called the Approximate Maximum Hit Ratio (AMHR). As a partial result, simulations show that hit ratios resulting from so-called mean ordered weights as surrogate weight vectors resemble the AMHR reasonably well. Further, we scrutinise the case of different distributions for the sampling of alternative values, with the result that it seems to be invariant, i.e. the general behaviour of surrogate weight methods does not change significantly. We also study the use of different filters on the AMHR. Finally, the corresponding already well-performing methods (e.g., ROC, RS, and SR) show good overall results. Thus, we conclude that the existing methods are fairly effective and provide decision-makers (individual or in groups) with a valuable means of efficiently eliciting and dealing with imprecise criteria information.

Keywords: Approximate maximum hit ratio · Multi-criteria decision analysis · Criteria weights · Criteria ranking · Distribution invariance · Performance limits · Surrogate weights

1 Introduction

Society is becoming more and more complex, year after year. As a consequence, decision-making also becomes more complicated, for individuals as well as teams. To aid humans in making better decisions, easy-to-use methods are one direction to follow. Using multi-criteria decision analysis (MCDA) is a well-established practice but sometimes puts too large a cognitive load on its users. To alleviate this burden, automatic weight generation methods (so-called surrogate weights) have been suggested. By a more natural elicitation process, the decision-makers understand the process better. This is also an advantage in group decision settings, when it could be complicated to arrive at numerical preferences but much easier to agree on an ordinal ranking, possibly after negotiations. Such negotiations are, again, facilitated by an easy-to-use elicitation

M. Campos Ferreira et al. (Eds.): GDN 2024, LNBIP 509, pp. 89–101, 2024.
https://doi.org/10.1007/978-3-031-59373-4_8

process. Within MCDA, the multi-attribute value theory (MAVT), also referred to as multi-attribute utility theory (MAUT), is the most widely used approach [1]. In order to rank the available alternatives, the additive model is the most extensively used [2], entailing that the resulting (expected) value is calculated according to Eq. (1). Another option would be the utilisation of a multiplicative model, which is described, for example, in [3]. This research assesses the outcomes obtained by employing the additive model, in which the utility of each alternative is quantified by Eq. (1).

$$V(a) = \sum\nolimits_{i=1}^{n} w_i v_i(a). \tag{1}$$

Using (1), the total sum $V(a)$, representing the overall utility for each alternative a, can be calculated by adding the products of w_{i} and v_{i} for all indices i. In this equation, w_{i} is the weight of criterion i out of n criteria, and v_{i} is the value of alternative a under criterion i. Ultimately, the alternative with the highest value is selected in this additive model.

One important issue with the application of MCDA models is that often, no precise quantitative information concerning the weights can be elicited, or the decision-maker (DM) does not wish to provide precise information [2]. This is especially the case when many criteria are being modelled [3, 4]. Several models have been proposed in the literature, such as second-order techniques [5, 6] and modifications to traditional decision rules [7, 8]. Additionally, some suggestions focus on ROC (Rank Order Centroid) methods or preference intensities [9–11]. If no precise information concerning the criteria weights can be assigned by the DM, ranking the criteria is a good way to elicit preference information. Using this ordinal information, surrogate weight methods can be applied to generate the weights. The elicitation of the weights can also be done in different ways. One method described later is based on the concept of point allocation, while the other described method is based on the concept of direct rating. These two methods admit different degrees of freedom (DoF) for the DM, which has a substantial impact on the performance of different surrogate weight models. For the performance of the models, the hit ratio is a standard measure of success [12]. Many different surrogate weight models have been proposed over the last decades [13]. Nevertheless, it has not been known what the maximum obtainable hit ratio for surrogate weight models is. Instead, different models were generally directly compared without considering an upper possible success limit. This could lead to a misallocation of research resources. Thus, we propose a simple approach and supposition that can easily approximate the maximum hit ratio in general situations, hence its name, the Approximate Maximum Hit Ratio (AMHR). Furthermore, we scrutinise the case of different distributions for sampling the alternative values. In previous research, the uniform distribution was often used, for example in [2]. Since an amount of information loss is suffered by the surrogate weight models, i.e., the precise but unknown underlying criteria weights for each decision situation are replaced by predetermined numbers (the surrogate weights of one model are the always the same for the specific rank since deterministic formulas generate them, see Section 2.1, only the number of used criteria influences the weights, i.e., w_{i} is the same for all alternatives), we wanted to find out whether the results of surrogate weight methods are sensitive to changes in the alternative value distributions.

Next, we will give an overview of rank ordering methods, the elicitation of weights, the construction of the AMHR, and special topics for surrogate weights modelling, namely alternative value distributions and filtering. Then, we show the testing results, followed by concluding remarks and areas for future research.

2 Rank Ordering Methods

Different techniques have been developed to cope with situations when the decision-maker (DM) is unable or does not wish to assign precise criteria weights. Surrogate weights, which are generated automatically, can be used when only the order of criteria weights is known but not their exact values. In this approach, the additive model is computed using surrogate weights derived from the imprecise information provided by the DM. These surrogate weights are intended to represent the DM's judgments regarding the decision situation [14–17].

Ordinal ranking is a key idea in relation to surrogate weights. In ordinal ranking, the criteria weights are ranked according to how important they are to the DM, but the distance (relative importance) between the weights is not required. In the standard approach, the order is given only by a simple order relation (>) between criteria. For example, Criterion 1 is more important than Criterion 2. In this paper, the following classic automatic generation methods are considered: Rank Sum weights (RS) [18], Rank Reciprocal weights (RR) [18], Rank Order Centroid weights (ROC) [19], Sum Rank (SR) weights [13], and Geometric Sum (GS) weights [13].

2.1 Ordinal Ranking

The classic methods RS and RR for ordinal ranking have been described extensively in [18]. Another classic method, ROC, was proposed in [19]. The Sum Rank (SR) method was proposed in [13]. It is a linear combination of the RS and RR methods to reduce the extreme behaviours of both methods. The SR method allocates the weights according to the formula

$$w_i^{SR} = \frac{1/i + \frac{N+1-i}{N}}{\sum_{j=1}^{N}\left(1/j + \frac{N+1-j}{N}\right)}, i = 1, \ldots, N; \sum w_i = 1; 0 \leq w_i. \tag{2}$$

In (2) and (3), the parameter N represents the total number of criteria. Additionally, we investigate the geometric sum (GS) method, which reflects the rank order multiplicatively in the numeric weights [13]. As seen in its formula below, GS contains a parameter s. This paper uses a value of 0.75 for the parameter, since it has been shown to be a good general choice [20]. For the geometric sum method, the formula to assign the criteria weights is

$$w_i^{GS} = \frac{s^{(i-1)}}{\sum_{j=1}^{N} s^{(j-1)}}, \ 0 < s < 1. \tag{3}$$

We direct readers to [13–19] for a more thorough treatment of the classic ordinal approaches well-known within the MCDA field.

2.2 Weight Elicitation

This paper discusses two primary methods of elicitation for determining criteria weights: point scoring and direct rating. The point allocation mechanism used in scoring points is based on [21]. In this approach, each criterion is assigned a specific number of points out of a predetermined total, such as 100 or 1000. The assigned point values correspond to the weights, with higher values indicating larger weights. The total point sum is calculated by adding up all the distributed points. The points for the last weight are determined based on the points given to the preceding N–1 weights, as the total point sum for N weights is fixed. Therefore, the point allocation method has N–1 degrees of freedom (DoF). On the other hand, in direct rating, arbitrary points or other metrics can be assigned to each criterion. These points are then normalised into weights by dividing each assigned point sum by the total point sum. This results in N degrees of freedom since the weights of all criteria add up to 1 and are independent of each other.

2.3 Approximate Maximum Hit Ratio

The search for good surrogate weight methods dates back a long time. For example, already Stillwell et al. studied the behaviour of equal weights serving as surrogate weights and compared them to RS and other methods [18]. The idea in this paper is to study the performance of different criteria weight vectors by exhaustively testing all possible vectors with regard to their performance. Since weights can be instantiated from a continuous range (between 0 and 1), we discretised the range of possible weights in an n-dimensional grid with 0.01 resolution for n criteria and tested all resulting possibilities, recording the hit ratios for each weight combination. For example, for $n = 3$ criteria, we generated, in steps of 0.01, weight vectors in the following sequence: (1, 0, 0), (0.99, 0.01, 0.00), (0.98, 0.02, 0.00), (0.98, 0.01, 0.01), (0.97, 0.03, 0.00), (0.34, 0.33, 0.33). This results in 884 possible weight combinations. For 4 criteria in steps of 0.01, this results in 8,037 possible weight combinations. Finally, for 5 criteria in steps of 0.02 (to keep the number of vectors down), this results in 3,765 possible weight combinations (instead of 46,262 if we had used 0.01 as the step). Each generated weight combination was used as a surrogate weight vector. In this manner, we can search for the criteria weight combination for each decision situation with the highest hit ratio, meaning that it points out the preferred alternative out of a set of alternatives with the highest success percentage. Since it is not possible to test all combinations due to the continuous nature of the weight vectors, this maximum hit ratio is approximate. Therefore, we call it the approximate maximum hit ratio (AMHR).

2.4 Different Distributions for the Alternative Values

In the literature, the uniform distribution is often used for the generation of alternative values (e.g., [7, 8, 20–23]). As mentioned in [8], already [12] studied whether using a normal distribution instead of a uniform distribution affects the hit ratios. It is obvious that linear scaling does not alter the results. E.g., using a Uniform(0, 1) distribution and a Uniform(0, 100) yield exactly the same results if the same seed settings are used.

However, in this study we look at whether a broader spectrum of value distributions significantly alters the resulting hit ratios. For that purpose, we use a lognormal distribution in order to test the possibility for more extreme values on one side of the distribution, and a Beta(0.5, 0.5) distribution, resulting in a U-shaped distribution, in order to test more extreme values on both sides of the distribution. In this way, we cover distributions far from what has hitherto been the standard in studies of surrogate weights.

2.5 Filtering

In real-life decision-making, the DM possibly refrains from using a surrogate weight method or similar methods if the weight vector is too "unnatural". There have been arguments against unrestricted surrogate methods, pointing out that "unnatural" weight distributions might occur in the weight vectors. This unnaturalness shows up as weights differ too much from each other [21]. A main remedy idea is to use some filter for the weight vectors that are deemed unnatural in some sense. We study the behaviour of different filters (0.1, 0.2, 0.3, and 0.4) on the AMHR. This means that if the largest and the second largest weights have a distance larger than, for example, 0.1, then the weight vector is resampled. The filter implementation in this study differs from [21] where weight vectors were discarded when a weight component was larger or smaller than certain thresholds.

2.6 Modelling Workflow

In the following, the general modelling workflow is described. In the beginning, the general simulation settings are determined: the seed, the number of runs, the number of criteria, the number of alternatives, the number of scenarios, and vectors containing parameters. Citing [12], this simulation verification method can be summarised as follows:
Step 1. Generate vectors at random from the weight space. These will be denoted as the TRUE weights.
Step 2. Generate an $m \times n$ matrix $\mathbf{V}$ of random values whose rows contain the attribute values associated with the ith alternative. By convention, the values in each column are scaled so that the smallest value is zero and the largest is one. This study deviates from the scaling approach in order to use different alternative value distributions.
Step 3. Calculate $\text{MAV}_i = \mathbf{v}_i{}^{\text{T}}\ \mathbf{w}$ for each of the m alternatives, substituting for $\mathbf{w}$, in turn, the weights for TRUE [and for the investigated surrogate weights, including the weight vectors for the AMHR evaluation]. Each choice of weights selects a best alternative (i.e., largest MAV).
Step 4. The selection decision under TRUE and the corresponding MAV are taken to be correct. The selections generated under the different weights are compared with that of TRUE.

The procedure above is the de facto standard tool for comparing surrogate weight methods. The idea of using simulations for comparisons was introduced in [12] and has been used by many researchers since, such as in [1, 7, 8]. The specific procedure in this paper follows [1] most closely. All of these versions of the simulation procedure generate well-defined datasets with known properties suitable for comparisons. An overarching idea, commonly accepted in the field, is that such a procedure well captures the intent of decision-makers as a cohort.

In order to generate the weights, both an N-generator and an N–1-generator were applied, i.e., methods that generate weights with N and N–1 DoF. For the N-generator, N independent values were sampled using a uniform distribution with normalisation of the generated independent values and for the N–1-generator a Dirichlet distribution was used. Thus, the generators are named after their freedom in generating random numbers.

The simulations in this study were conducted using R, a statistical software that is open source. The simulation was run a large number of times until the results converged. For each run, we compared whether each method had the same best alternative as the one indicated by the TRUE weights. In case of a match, a hit (1) was recorded. If no match is obtained, then (0) was recorded as the result, meaning no hit. The hit ratio is then calculated as the number of hits for each method divided by the overall number of rounds in a particular simulation.

3 Results

3.1 Comparison of Different Distributions for the Alternative Values

For the comparison of different alternative value distributions, the combinations of 3 criteria and 3 alternatives, 6 criteria and 6 alternatives and 9 criteria and 9 alternatives are compared using three different alternative value distributions. Table 1 shows the comparison of the above-mentioned surrogate weight methods using a Uniform(0, 1), a Beta(0.5, 0.5) and a Lognormal(0, 1) distribution. The DoF split columns show the mean of the hit ratios over a 50/50 mix of the N and N–1 DoF distributions. This gives information about the robustness of the studied methods since we cannot know the decision-maker's reasoning in any detail when we elicit the data. A split of 1.0 means that only the N-generator was used, whereas a split of 0.0 corresponds to using only the N–1-generator and 0.5 denotes an equal mix of both generators. The 0.5 DoF split is used to show the performance when a mix of DoF is used, i.e. when the particular reasoning of the DM is not known (which in practice is almost always). For each combination, 100,000 decision situations were generated. The best-performing methods for each DoF setting are marked in green.

The difference between the various value distributions is indeed small. This is important to note since it indicates that surrogate weights are relatively independent of the underlying value distributions. Since it is often hard to determine value distributions by elicitation methods in real life, this is an encouraging result for the widespread use of surrogate weights. Furthermore, the performance of RR, RS, ROC, GS, and SR are in accordance with previous studies [13, 20, 22].

Table 1. Hit ratios using a Uniform(0, 1), a Beta(0.5, 0.5), and a Lognormal(0, 1) distribution. The columns represent an N–1-generator (left), an N-generator (right) and an equal mix (middle)

		Uniform			Beta			Log		
3x3		0	0.5	1	0	0.5	1	0	0.5	1
	ROC	89.6	89.1	88.0	90.1	88.7	88.2	90.2	89.9	89.6
	RS	87.7	89.3	90.4	88.2	88.9	90.5	88.8	90.1	91.6
	RR	88.3	89.2	89.2	88.8	88.7	89.1	88.9	89.5	90.6
	SR	88.3	89.5	90.0	88.6	89.2	90.2	89.0	90.1	91.3
	GS	82.9	86.4	88.5	83.9	86.1	88.6	85.4	87.4	90.1
6x6										
	ROC	84.8	82.4	80.1	85.7	83.2	81.1	85.9	84.8	81.9
	RS	80.1	83.7	87.2	80.8	84.0	87.5	81.0	85.7	88.6
	RR	82.6	80.4	77.9	83.3	81.1	79.1	83.8	82.7	80.0
	SR	83.1	84.2	84.9	83.8	84.5	85.6	84.0	86.0	86.5
	GS	81.0	83.5	85.8	81.7	83.8	86.3	81.9	85.4	87.3
9x9										
	ROC	83.7	80.2	76.4	84.1	80.6	76.8	84.7	81.5	77.6
	RS	76.1	81.5	87.4	76.9	82.0	87.3	77.1	83.2	88.6
	RR	79.7	74.9	70.2	80.3	75.8	71.2	80.7	76.0	70.6
	SR	81.4	82.3	83.2	81.8	82.6	83.2	82.6	83.8	84.4
	GS	83.2	82.0	80.8	83.5	82.4	80.8	84.2	83.3	82.1

3.2 Mean Ordered Weights

After observing the behaviours of the different alternative values distributions, we study the following interesting behaviour, namely the difference in the generators as described in [19]. Therefore, to better understand the criteria weights generated from the different generators, we simulated for each weight generator 1,000,000 weight vectors. Then, the weights were ordered, and the mean values of the ordered weights were calculated. This was done to obtain insights about the general shape of simulated weight vectors. Since ROC and RR perform better for N–1 DoF, and knowing that ROC, as described above, results in more extreme weights, we assumed that the ordered weights would show a more extreme pattern.

Table 2. Mean ordered weights for 3, 4, and 5 criteria using N and N–1 DoF

	w1	w2	w3	w1	w2	w3	w4	w1	w2	w3	w4	w5
N	0.52	0.32	0.15	0.42	0.30	0.19	0.09	0.35	0.27	0.20	0.13	0.06
N–1	0.61	0.28	0.11	0.52	0.27	0.15	0.06	0.46	0.26	0.16	0.09	0.04

We can see that for N–1 DoF, the mean weights of the most important criterion are generally higher than for N DoF. This is in accordance with previous literature [1], which mentions that ROC and RR, who perform better for the N–1 DoF case, yield more extreme weights. Especially ROC, which is aligned only with this elicitation case, has been criticised in that respect. Using mean ordered weights, we can further investigate how this relates to the best-performing surrogate weight methods.

3.3 Approximate Maximum Hit Ratio

For all AMHR simulations, 10,000 decision situations were simulated for each vector. On the left in Tables 3 and 4, the hit ratios for the best-performing criteria weight vectors for N–1 DoF are shown, and on the right for N DoF (HR means hit ratio). The most important criterion is c1 in all cases.

Table 3. AMHR for 3 criteria and 3 alternatives (884 combinations)

Rank	c1	c2	c3	N–1 HR	Rank	c1	c2	c3	N HR
1	0.58	0.31	0.11	90.23	1	0.49	0.33	0.18	89.34
2	0.61	0.28	0.11	90.22	2	0.48	0.33	0.19	89.33
3	0.60	0.28	0.12	90.22	3	0.48	0.34	0.18	89.27
4	0.59	0.29	0.12	90.22	4	0.47	0.34	0.19	89.23
5	0.58	0.29	0.13	90.22	5	0.50	0.34	0.16	89.20
					...	...	...	...	...
					27	0.52	0.33	0.15	88.96

Table 4. AMHR for 4 and 5 criteria and 3 alternatives (8037 and 3765 combinations)

Rank	c1	c2	c3	c4		N–1 HR	Rank	c1	c2	c3	c4		N HR
1	0.52	0.28	0.13	0.07		90.87	1	0.39	0.32	0.20	0.09		90.46
2	0.51	0.28	0.15	0.06		90.84	2	0.38	0.31	0.20	0.11		90.40
3	0.51	0.28	0.14	0.07		90.84	3	0.39	0.31	0.20	0.10		90.39
4	0.51	0.27	0.15	0.07		90.84	4	0.38	0.32	0.20	0.10		90.39
5	0.51	0.27	0.14	0.08		90.83	5	0.40	0.30	0.20	0.10		90.38
...	...	...	...	...		...	...	...	...	...	...		...
30	0.52	0.27	0.15	0.06		90.69	36	0.42	0.30	0.19	0.09		90.21
Rank	c1	c2	c3	c4	c5	N–1 HR	Rank	c1	c2	c3	c4	c5	N HR
1	0.42	0.26	0.16	0.10	0.06	89.64	1	0.34	0.26	0.20	0.14	0.06	91.17
2	0.44	0.26	0.16	0.10	0.04	89.56	2	0.34	0.28	0.20	0.12	0.06	91.06
3	0.44	0.24	0.18	0.10	0.04	89.54	3	0.32	0.26	0.20	0.14	0.08	90.95
4	0.46	0.26	0.16	0.08	0.04	89.51	4	0.34	0.26	0.18	0.16	0.06	90.91
5	0.44	0.24	0.16	0.10	0.06	89.50	5	0.34	0.28	0.20	0.14	0.04	90.88

It can be seen that the best combinations are often quite close to each other, i.e. they converge to discrete weight distributions that seem to have some general properties. Marked in green are weight combinations closest to the mean ordered weights. It can be seen that the mean ordered weights generally perform as a good approximation of the AMHR.

3.4 Applying Filters for the AMHR

A common criticism of the simulation method used in this study, as well as in many others, is that the random generator can produce random vectors that would be highly unlikely to occur in real-life situations. For instance, a vector of three weights (0.9, 0.05, 0.05). Although nothing is formally incorrect about the vector, and the simulations themselves run smoothly, a decision-maker would likely not utilise a multi-criteria decision method if one criterion was overwhelmingly dominant.

To address this issue, filters were applied to the random vector generation process, which excluded vectors with the property described in Sect. 2.5 from participating in the simulations. This led to a new set of results, as shown in Table 5, which present the results for decision scenarios involving three criteria. Similar patterns were observed for other problem sizes.

We can see that using a filter has a visible effect on the AMHR, especially for the lower filter thresholds 0.1 and 0.2. As described in Sect. 2.5, a filter of 0.1 means that a weight vector is resampled if the distance between the largest and the second largest weight is more than 0.1. One might argue that, for example, 0.1 is an unreasonably low upper limit for the difference between the two most important criteria, but there are a lot of such decision situations in real life. It can be observed that there is a difference in the best-performing weight vectors between N and N–1 DoF, with N DoF resulting in a generally lower weight for the most important criterion.

Table 5. Applying filters to the simulations for N and N–1 DoF

Filter					N–1				N
	Rank	c1	c2	c3	HR	c1	c2	c3	HR
10%									
	1	0.45	0.40	0.15	94.14	0.43	0.37	0.20	92.85
	2	0.45	0.39	0.16	94.10	0.42	0.37	0.21	92.79
	3	0.46	0.40	0.14	94.09	0.43	0.38	0.19	92.73
20%									
	1	0.48	0.38	0.14	93.63	0.45	0.35	0.20	92.13
	2	0.49	0.38	0.13	93.60	0.46	0.36	0.18	92.07
	3	0.49	0.37	0.14	93.60	0.45	0.34	0.21	92.06
30%									
	1	0.50	0.38	0.12	91.86	0.46	0.36	0.18	93.15
	2	0.51	0.36	0.13	91.83	0.45	0.36	0.19	93.10
	3	0.52	0.37	0.11	91.82	0.45	0.35	0.20	93.06
40%									
	1	0.53	0.33	0.14	91.01	0.49	0.35	0.16	91.41
	2	0.54	0.34	0.12	90.99	0.48	0.36	0.16	91.40
	3	0.54	0.33	0.13	90.99	0.47	0.35	0.18	91.39

3.5 AMHR for Different Distributions for the Alternative Values

To emphasise the results in Sect. 3.1, the AMHR was computed using the two value-generating functions (other than the uniform) from that section. This section shows only the case of 5 criteria and 3 alternatives, but all results were similar. As expected, the results in Tables 6 and 7 are similar to those in Table 4, reinforcing the result that the uniform distribution is a very good candidate when selecting generating functions for alternative values in surrogate simulation studies since it can cover a wide range of distributions and in real-life decision situations, the value distributions are seldom known.

Table 6. AMHR for different distributions for the alternative values – Beta(0.5, 0.5)

Rank	c1	c2	c3	c4	c5	N−1 HR	Rank	c1	c2	c3	c4	c5	N HR
1	0.44	0.28	0.14	0.1	0.04	89.39	1	0.34	0.26	0.2	0.14	0.06	91.06
2	0.44	0.28	0.16	0.08	0.04	89.36	2	0.32	0.28	0.2	0.14	0.06	90.93
3	0.44	0.26	0.16	0.1	0.04	89.36	3	0.34	0.26	0.2	0.12	0.08	90.88
4	0.42	0.28	0.16	0.1	0.04	89.33	4	0.36	0.26	0.2	0.12	0.06	90.86
5	0.46	0.26	0.14	0.1	0.04	89.28	5	0.34	0.26	0.22	0.12	0.06	90.86

Table 7. AMHR for different distributions for the alternative values – Lognormal(0, 1)

Rank	c1	c2	c3	c4	c5	N−1 HR	Rank	c1	c2	c3	c4	c5	N HR
1	0.44	0.26	0.16	0.1	0.04	89.68	1	0.34	0.26	0.2	0.14	0.06	92.50
2	0.46	0.26	0.16	0.08	0.04	89.63	2	0.32	0.28	0.2	0.14	0.06	92.37
3	0.48	0.26	0.14	0.08	0.04	89.62	3	0.34	0.28	0.18	0.14	0.06	92.33
4	0.46	0.26	0.14	0.1	0.04	89.61	4	0.36	0.26	0.18	0.14	0.06	92.31
5	0.44	0.28	0.16	0.08	0.04	89.59	5	0.34	0.28	0.2	0.12	0.06	92.31

3.6 Minimum Hit Ratios for Ordered Weight Factors

The simulations in Sect. 3.3, show that the worst hit ratios for the generated ordered weight vectors are, in most cases, when almost all weight is given to the most important criterion. That is the case for both DoFs. Nevertheless, the hit ratio is always clearly above 60%. That is, as long as some ranking is used, this is beneficial to the DM compared to mere guessing or equal weights.

4 Concluding Remarks

The utilisation of surrogate weights for decision-analytic support has become more prevalent in recent years, in attempts to let the decision-makers focus more on content and less on formalities of elicitation, thus putting the decision-makers more in control of the decision process. Consequently, there has been an increasing need to explore a broader range of potential weight methods. The findings of this study indicate that the hit ratios resulting from the mean ordered weights as surrogate weight vectors closely resemble the AMHR. However, it should be noted that only a limited number of decision situations were tested. Therefore, whether the AMHR can be generally approximated for other decision scenarios remains to be established. We suggest, in order to test the generalizability of the approximation of the AMHR through the mean ordered weights, to widen the combinations of possible decision situations. However, the difference in the hit ratios resulting through different DoF is the main obstacle in finding a one-fits-all approximation with the current model parameters. Nonetheless, the existing methods provide decision-makers with a valuable means of dealing with imprecise criteria information. Furthermore, especially for the pure N and N–1 DoF, the corresponding already well-performing methods (e.g., ROC for N–1 DoF and RS for N DoF) show good overall results. Nevertheless, the difference between N and N–1 DoF cannot, by definition, be eliminated. I.e., no surrogate method can be close to the optimal hit ratio for both DoF. Interestingly, the AMHR can be somewhat increased by filtering "unnatural" decision situations. Additionally, the use of more extreme value distributions has only minor effects and the results for DoF for the criteria weight creation are in accordance with previous findings. Altogether, the DoF impacts the model more than the alternative values distributions.

4.1 Discussion

This paper explores the upper limit of possible hit ratios for surrogate weight methods. This was conducted by broadly testing possible weight vector combinations, resulting in the AMHR. According to the authors' knowledge, this is the first time such a limit has been established. This finding will help in directing surrogate weight research in the future. When methods that are very close to the AMHR limit have been found, efforts could be directed elsewhere. Further, we found that the mean ordered weights as surrogate weight vectors result in hit ratios that resemble the AMHR quite well. The number of decision situations studied is comparatively small, so a generalisation is not yet possible. Additionally, we studied the use of different distributions for the alternative values, resulting in the conclusion that the general behaviour of widely used surrogate weight methods does not change significantly. Nevertheless, the gap between N and N–1 DoF is still significant regardless of the used distribution for the alternative values. Hence, the DoF clearly seems to be more important for the final hit ratio than the alternative value distribution used.

5 Further Research

There are many possibilities for further studies of surrogate weight methods. For example, evaluating whether the AMHR fits other situations, like filtering extreme events as in [19], or fine-tuning the AMHR. Furthermore, it would be interesting to see whether the AMHR can be expanded to cardinal ranking (taking distances between elements in orderings into account), where the generation of the weights is not as straightforward as for the ordinal case. Future research could also consider whether the mean ordered weights perform similarly well in other decision situations and what a suitable trade-off for the DMs' varying and most often unknown DoFs could be. With the insights from the AMHR, future research can use these results as benchmarks in order to rank new models, especially concerning performance and simplicity. It is also possible to direct research resources in other directions when results close to the limit have been obtained, and there is not much leverage left to discover.

Acknowledgements. This paper is dedicated to the co-author, dear friend, and esteemed colleague Professor Love Ekenberg, who passed away in September 2022 during the research leading up to this paper.

Disclosure of Interests. The authors have no competing interests to declare relevant to this article's content.

References

1. Danielson, M., Ekenberg, L.: A robustness study of state-of-the-art surrogate weights for MCDM. Group Decis. Negot. **26**, pp. 677–691 (2016)
2. Aguayo, E.A., Mateos, A., Jiménez, A.: A new dominance intensity method to deal with ordinal information about a DM's preferences within MAVT. Knowl.-Based Syst. **69**, pp. 159–169 (2014)
3. Park, K.S.: Mathematical programming models for characterizing dominance and potential optimality when multicriteria alternative values and weights are simultaneously incomplete. IEEE Trans. Syst. Man Cybern – Part A: Syst. Hum. **34**(5), pp. 601–614 (2004). https://doi.org/10.1109/TSMCA.2004.832828
4. Larsson, A., Riabacke, M., Danielson, M., Ekenberg, L.: Cardinal and rank ordering of criteria – addressing prescription within weight elicitation. Int. J. Inf. Technol. Decis. Mak. **14**(6), pp. 1299–1330 (2014)
5. Danielson, M., Ekenberg, L.: Computing upper and lower bounds in interval decision trees. Eur. J. Oper. Res. **181**(2), pp. 808–816 (2007)
6. Ekenberg, L., Danielson, M., Larsson, A., Sundgren, D.: Second-order risk constraints in decision analysis. Axioms **3**(1), pp. 31–45 (2014)
7. Ahn, B.S., Park, K.S.: Comparing methods for multiattribute decision making with ordinal weights. Comput. Oper. Res. **35**(5), pp. 1660–1670 (2008)
8. Sarabando, P., Dias, L.: Multi-attribute choice with ordinal information: a comparison of different decision rules. IEEE Trans. Syst. Man Cybern. Part A **39**, pp. 545–554 (2009)
9. Bana e Costa, C.A., Correa, E.C., De Corte, J.M., Vansnick, J.C.: Facilitating bid evaluation in public call for tenders: a socio-technical approach. Omega **30**, pp. 227–242 (2002)
10. Sarabando, P., Dias, L.: Simple procedures of choice in multicriteria problems without precise information about the alternatives' values. Comput. Oper. Res. **37**, pp. 2239–2247 (2010)
11. Figueira, J., Roy, B.: Determining the weights of criteria in the ELECTRE type methods with a revised Simos' procedure. Eur. J. Oper. Res. **139**, pp. 317–326 (2002)
12. Barron, F., Barrett, B.: Decision quality using ranked attribute weights. Manage. Sci. **42**(11), pp. 1515–1523 (1996)
13. Danielson, M., Ekenberg, L.: Rank ordering methods for multi-criteria decisions. In: Zaraté, P., Kersten, G.E., Hernández, J.E. (eds.) Group Decision and Negotiation. A Process-Oriented View. GDN 2014. Lecture Notes in Business Information Processing, vol. 180, pp. 128–135. Springer, Cham (2014). https://doi.org/10.1007/978-3-319-07179-4
14. Arbel, A., Vargas, L.G.: Preference simulation and preference programming: robustness issues in priority derivation. Eur. J. Oper. Res. **69**, pp. 200–209 (1993)

15. Barron, F., Barrett, B.: The efficacy of smarter: simple multi-attribute rating technique extended to ranking. Acta Physiol (Oxf.) **93**(1–3), pp. 23–36 (1996)
16. Katsikopoulos, K., Fasolo, B.: New tools for decision analysis. IEEE Trans. Syst. Man, Cybern. – Part A: Syst. Hum. **36**(5), pp. 960–967 (2006)
17. Stewart, T.J.: Use of piecewise linear value functions in interactive multicriteria decision support: a monte carlo study. Manage. Sci. **39**(11), pp. 1369–1381 (1993)
18. Stillwell, W., Seaver, D., Edwards, W.: A comparison of weight approximation techniques in multiattribute utility decision making. Organ. Behav. Hum. Perform. **28**(1), pp. 62–77 (1981)
19. Barron, F.H.: Selecting a best multiattribute alternative with partial information about attribute weights. Acta Physiol (Oxf.) **80**(1–3), pp. 91–103 (1992)
20. Lakmayer, S., Danielson, M., Ekenberg, L.: Aspects of ranking algorithms in multi-criteria decision support systems. In: Fujita, H., Guizzi, G. (eds.) New Trends in Software Methodologies, Tools and Techniques, pp. 63–75. IOS Press, Amsterdam (2023)
21. Danielson, M., Ekenberg, L.: The CAR method for using preference strength in multi-criteria decision making. Group Decis. Negot. **25**, pp. 775–797 (2016)
22. Lakmayer, S., Danielson, M., Ekenberg, L.: Automatically generated weight methods for human and machine decision-making. In: Fujita, H., Wang, Y., Xiao, Y., Moonis, A. (eds.) Advances and Trends in Artificial Intelligence. Theory and Applications. IEA/AIE 2023. Lecture Notes in Computer Science, vol. 13925, pp. 195–206. Springer, Cham (2023) https://doi.org/10.1007/978-3-031-36819-6_17
23. Salo, A.A., Hämäläinen, R.P.: Preference ratios in multiattribute evaluation (PRIME)-elicitation and decision procedures under incomplete information. IEEE Trans. Syst. Man, Cybern – Part A: Syst. Hum. **31**(6), pp. 533–545 (2001)

Evaluation of the Degree of Manipulability of Positional Aggregation Procedures in a Dynamic Voting Model

Daniel Karabekyan[1] (✉) and Viacheslav Yakuba[1,2]

[1] HSE University, Moscow 101000, Russia
dkarabekyan@hse.ru

[2] Institute of Control Sciences of Russian Academy of Sciences, Moscow 117997, Russia

Abstract. The degree of individual manipulability of positional aggregation procedures is evaluated for the dynamic voting model within the framework of the two-dimensional Downsian model. In the dynamic voting model, alternatives move in steps toward a central point while agents attempt to manipulate at each step. The share of manipulable profiles, i.e., the Nitzan-Kelly index, is calculated. The manipulability of Plurality, Approval, Inverse Plurality, Borda, Threshold, Nanson, Inverse Borda, Hare, and Coombs aggregation procedures is evaluated at each step of the movement. Computer simulation shows that the Hare, Coombs, and Threshold procedures are among the least manipulable ones.

Keywords: Manipulability · Aggregation Procedures · Nitzan-Kely index · Dynamic Voting

1 Introduction

In [1, 2] a dynamic voting model was proposed and studied. The problem statement is as follows. We consider m alternatives and n agents located in two-dimensional space in accordance with the Downsian model [3]. According to the Downsian model, the alternatives and agents are placed on the plane. So the agents and the alternatives are assigned two coordinates. The winner is the alternative chosen according to a fixed decision-making procedure. In our case, positional procedures are considered. The losing alternatives successively move towards the so-called "winning rose" in order to get closer to the agents' ideal points than the competing alternatives and thus win, but without moving too far away at each step from the original position of the alternative. The "winning rose" concept is introduced in [1, 2], it highlights the fact, that the area of the winning positions of the alternatives, resembles the image of petals on the plane. As a result of such movement, dynamic trajectories are formed. This paper examines the degree of manipulability of positional decision-making procedures in the process of such step-by-step movement of alternatives with fixed positions of agents. The analysis is carried out using computer simulation using the Monte Carlo method. Random voting profiles are generated as the positions of agents and alternatives on the plane. The manipulability of both the original profile and all profiles on the dynamic path is evaluated.

M. Campos Ferreira et al. (Eds.): GDN 2024, LNBIP 509, pp. 102–113, 2024.
https://doi.org/10.1007/978-3-031-59373-4_9

2 Manipulability in the Two-Criteria Downsian Model

Computer simulations of the degree of manipulability in a dynamic voting model are performed for the following positional decision-making procedures: Plurality, Approval $q = 2, 3$, Inverse Plurality, Borda, Threshold, Nanson, Inverse Borda, Hare, and Coombs.

Definitions of the procedures considered are given in [4]. The evaluation of manipulability is performed for the case of multiple choice; the choice can be either single alternative or several alternatives. To allow comparison of choice results, extended preferences were introduced [5], and the discussion of extended preference models is also given in [6, 7].

In the two-dimensional Downsian model [3], m alternatives ($m = 3, 4$, or 5) and n ideal points of agents ($n > 2$) are located on a unit square on the coordinate plane. The agent's preferences are constructed as an ordering of alternatives by the Euclidean distance from each alternative to the agent's ideal point, i.e., alternatives are ordered by proximity to the agent's ideal point. The closest alternative takes the first place in the agent's preferences, the next one takes the second place, etc., and the most distant alternative takes the last place.

Manipulation is performed by the agent individually by shifting its ideal point in such a way that, by presenting insincere preferences, the agent obtains a better result of the social choice for herself compared to sincere preferences. The agent can move its position only within the allowed area – a square on the plane. The manipulability analysis in the two-dimensional Downsian model is performed in [8].

Let us consider an example of manipulation for 3 alternatives, 4 agents, for the Risk-averse (PBest) extension of the preferences for the Borda rule. Let the ideal points of the agents Ag1-Ag4 and alternatives a, b, and c be located on the plane, as shown in Fig. 1.

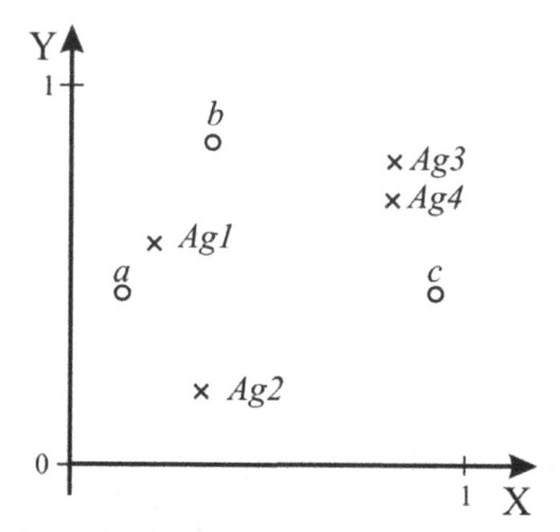

Fig. 1. Positions of ideal points of 4 agents and 3 alternatives in a two-dimensional Downs model

The extended preferences are based on the original preferences of the agent, like $a \succ b \succ c$ The extensions are used, if the result of the collective choice consists of sets of alternatives. The PBest extension for 3 alternatives is defined as: $\{a\} \succ \{a, b\} \succ \{b\} \succ \{a, b, c\} \succ \{a, c\} \succ \{b, c\} \succ \{c\}$.

Let us construct a profile of agents' preferences for the example. For each agent, the ordering of alternatives is constructed according to the proximity to the agent's ideal point. For example, for agent Ag1, the ordering is as follows: $a \succ b \succ c$.

The preference profile is given in Table 1.

Table 1. Example preferences

Ag1	Ag2	Ag3	Ag4
a	a	c	c
b	b	b	b
c	c	a	a

The choice according to the Borda rule is {a,b,c}.

If the agent Ag1 manipulates by shifting its position towards alternative b, as shown in Fig. 2, then the agent presents the following ordering of alternatives: $b \succ a \succ c$. The positions of the remaining agents do not change.

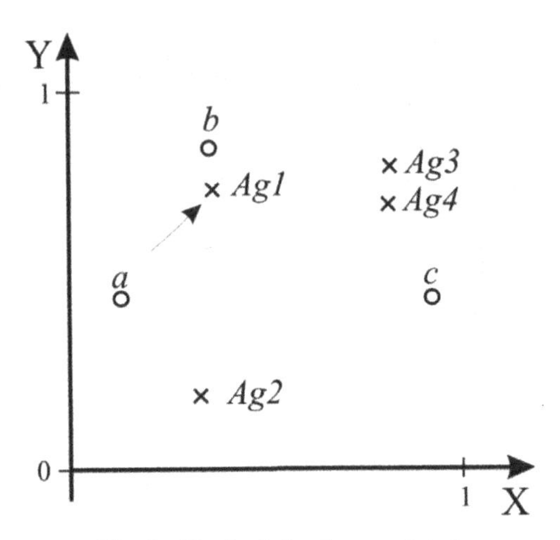

Fig. 2. Manipulation by agent Ag1

The profile after manipulation by agent Ag1 is shown in Table 2.

The choice by the Borda rule after manipulation by agent Ag1 is equal to {b}.

According to the Risk-averse (PBest) extended preferences for 3 alternatives, for agent Ag1 the choice {b} is more preferable to choice {a,b,c}. Thus, the social choice resulting from the manipulation of agent Ag1 is more preferable for her than with sincere preferences, and the profile in question is manipulable.

Table 2. Example of the manipulated profile

Ag1	Ag2	Ag3	Ag4
b	a	c	c
a	b	b	b
c	c	a	a

Table 3. Dynamics of the NK index, steps s1–s5, Leximin extension, 3 alternatives, 25 agents

	Spa	s1	s2	s3	s4	s5
Plurality	0.05	0.07	0.11	0.15	0.21	0.26
Approval q = 2	0.06	0.07	0.09	0.12	0.14	0.14
Inverse Plurality	0.06	0.07	0.09	0.12	0.14	0.14
Borda	0.03	0.03	0.05	0.07	0.11	0.18
Threshold	0.04	0.05	0.06	0.08	0.10	0.11
Nanson	0.03	0.03	0.05	0.07	0.11	0.18
Inverse Borda	0.03	0.03	0.05	0.07	0.11	0.18
Hare	0.01	0.01	0.02	0.04	0.07	0.11
Coombs	0.02	0.03	0.04	0.06	0.08	0.11

3 Manipulation in a Dynamic Voting Model

Let us consider the dynamic voting model proposed in [2, 3, 9, 10]. We consider m alternatives and n agents located in the two-dimensional space in accordance with the two-criteria Downsian model. The winner is the alternative chosen according to the aggregation procedure. The alternatives move towards the so-called "winning rose" in order to get closer to the agents of the competing alternatives and thus win. As a result of such movement, dynamic trajectories are formed. The properties of the trajectories are studied in [1, 2].

In the above mentioned papers, a two-player game is proposed. Each player places her own alternative on the two-dimensional space, and the winner of the two is determined by the simple majority rule. The position of the losing alternative is changed, and the game is repeated. Based on this model, we aim to evaluate the manipulability of several aggregation procedures in the situation that somehow reflects the very idea of the original model – the movement of the alternatives, step by step, to a midpoint of the simple majority of the agents, closest to the initial position of the alternative. Since, if we take the midpoint of the simple majority of the agents, the trajectory is the same for all considered aggregation rules. This looks like one of the simplest ways to compare the manipulability of the rules. It is, however, noticed that using the midpoint of the winning coalition for each of the rules might provide a more justified result. This reserves the room for further study.

We also need to clarify the motivations of the agents. In this model, the agents do not plan to construct the path for the alternatives to the final point. They pursue the nearest aim only, to obtain a better result of the collective choice through manipulation. So, for the agents, the positions of the alternatives are fixed. A particular agent can see that by presenting insincere preferences over alternatives, she can obtain a better result of the collective choice. But the ideal points of the agents do not move; the agents just pretend to be moving to manipulate the result of the aggregation procedures.

The main goal of this paper is to evaluate how the manipulability of the aggregation procedures themselves changes while the alternatives move to the more competitive position, while agents do not move but just manipulate, presenting insincere preferences.

So, we examine the degree of manipulability of positional decision-making procedures in the process of such step-by-step movement of alternatives with fixed positions of agents. Namely, the points formed are M_i – midpoints for the ideal positions of coalitions of the simple majority of agents closest to the alternative, as shown in Fig. 3.

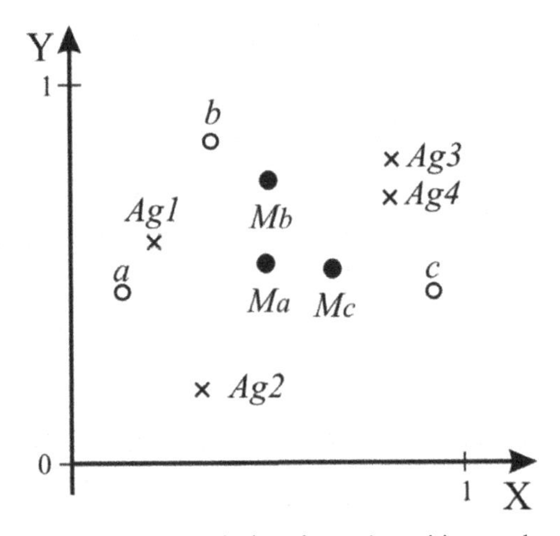

Fig. 3. Midpoints of simple majority of agent's positions and voting profile

The alternatives are then moved, by steps, towards their midpoints M_i as shown in Fig. 4.

For calculation time reasons, the number of steps from the alternative i to the respective M_i point is fixed at five. In the process of shifting alternatives, their ordering by proximity to agents changes, and so does the voting profile. The question arises: how does the manipulability of the procedures change in the process of movement of the alternatives towards the mid-points?

Random voting profiles are generated using the Monte Carlo method. The ideal positions of agents and alternatives are formed on a unit square on the plane.

The degree of manipulability of the procedures is calculated by sequentially changing the coordinates of the alternatives along the straight segment from the initial positions of the alternatives to the midpoints of the coalitions of the simple majority of agents closest to the alternatives in each of the randomly generated profiles. The segment is divided

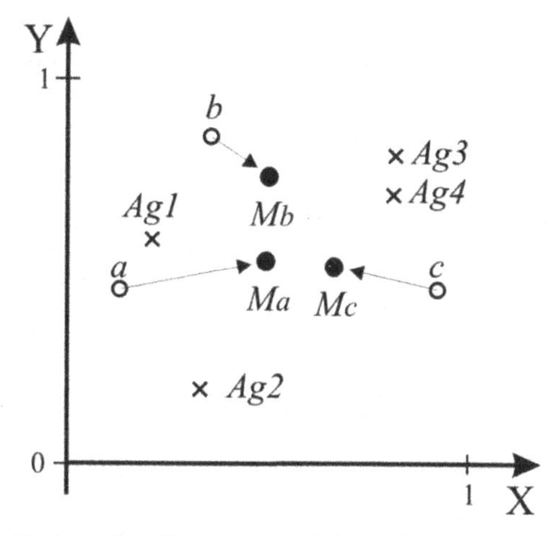

Fig. 4. Trajectories of movement of alternatives to the midpoints.

into five steps. For each combination of the alternatives and agents (m = 3,4,5 and n = 3–25), 10000 random profiles are generated. For each constructed profile, segments are formed from the initial position of each alternative to the midpoint of the coalition of the simple majority of agents closest to this alternative. On the segment constructed in this way, k = 5 points s1,...,s5 are specified, dividing the segment into k + 1 equal parts. The point for the original profile is denoted as Spa. Each of the alternatives is moved along the corresponding segments to points s1,..., s5 consequently. For each step, for each alternative, the degree of manipulability, as a share of the manipulable profiles, is calculated. This degree is then averaged over that obtained for all alternatives on the current step.

In this dynamic model, as well as in the original static model of manipulability in the Downsian framework, the Euclidean distance is used. Tie-break situations and degenerate cases are not considered, since the probability of their occurrence is close to zero.

4 Calculation Scheme for the Evaluation of the Manipulability

Let us present a scheme for evaluating the manipulability of the procedures. On the two-criteria plane, a square with side 1 is specified, laid out from the origin. A profile is randomly generated from m alternatives (m = 3, 4, 5) and n agents (n = 3–25) with X and Y coordinates from 0 to 1 in accordance with a uniform distribution. Without loss of generality, it is assumed that in profiles the two most distant alternatives, for example, a and e in the case of 5 alternatives, are placed in fixed positions a[0,0], e[1,0], and all alternatives are ordered alphabetically with increasing coordinate X. These particular positions are just taken for convenience. Other rescaling options are possible, e.g., a[0,0], e[1,1], etc.

The orderings of the agents' preferences are constructed according to the proximity of the alternatives to the agent's ideal point in accordance with the Euclidean distance. The manipulation scheme is based on the static model studied in [8]. The agent manipulates individually, and the insincere preferences are constructed by shifting the agent's ideal point along the X and Y coordinates on a grid with steps of 1/L. The parameter L can be varied to balance the calculation resources needed and the accuracy of the result. The results in this paper are obtained for L = 100. The agent's preferences are then constructed by ordering alternatives by the Euclidean distance to this point.

The choices are calculated by each of the aggregation procedures independently. The choice results obtained from sincere preferences and from insincere preferences are compared using all versions of extended preferences. If at least one of the agents benefits from the manipulation, i.e., for her preferences, the choice based on the insincere preferences is better than the choice based on the sincere preferences in accordance with a particular preference extension, then the manipulation is considered successful for a given profile, procedure, and the type of extended preference.

In this work, only individual manipulation of the agents is considered, i.e., each agent presents insincere preferences alone, while other agents do not try to manipulate. It is certainly possible to consider the models in which coalitional manipulability is used with different methods to form a coalition. This might lead to interesting results in further research.

For each procedure and preference extension, the number of manipulated profiles is counted, and the Nitzan-Kelly manipulability index [11, 12] is calculated as the ratio of the number of manipulable profiles to the total number of profiles considered. The index is defined as follows:

$$NK = \frac{d_0}{d_{total}} \tag{1}$$

where d_0 - number of manipulated profiles, d_{total} - total number of profiles.

5 Results

The dynamics of the NK index differ for different aggregation procedures. For 3 alternatives, rules such as Plurality, Inverse Borda, and Nanson exhibit a significant increase in manipulability when moving from the starting point Spa to point s5. Plurality becomes the most manipulable for almost all numbers of agents and for all extensions.

The least manipulable at the starting point Spa is the Hare rule, maintaining a low degree of manipulability along the trajectory to s5. Although the manipulability of all procedures increases, the Threshold and Coombs rules show a slowdown in the growth of manipulability, and by s5, the degrees of manipulability for these two rules become close Fig. 5.

For 4 alternatives, at the initial stage of the trajectory, with a sufficiently large number of agents, the Coombs and Hare procedures are the least manipulated. However, for these two procedures, as well as for Inverse Borda, Nanson, Plurality, and Approval q = 2, manipulability increases as they move towards point s5. On the contrary, for the Threshold, Approval q = 3, and Inverse Plurality rules, manipulability decreases, with the manipulability of the Threshold rule at point s5 becoming less than that of the Coombs rule. The manipulability of the Approval q = 3 and Inverse Plurality rules,

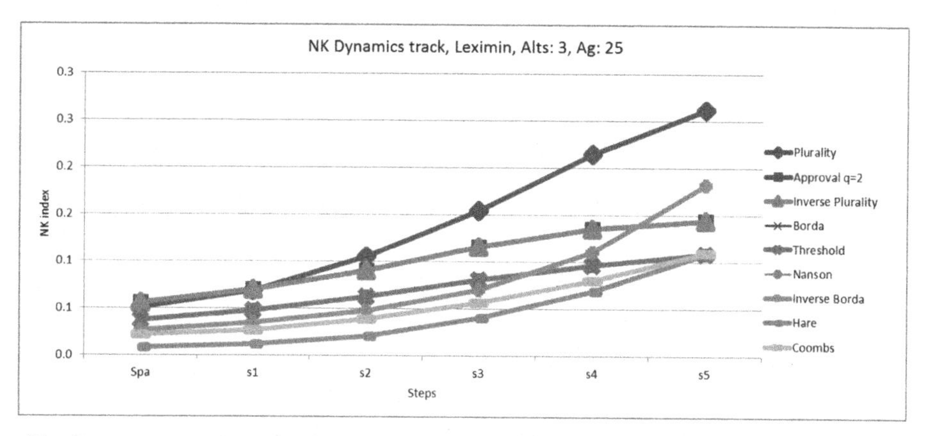

Fig. 5. Dynamics of the NK index, steps s1–s5, Leximin extension, 3 alternatives, 25 agents

which is greater than that of the Coombs rule in s5, becomes less than the manipulability of the Hare rule Fig. 6.

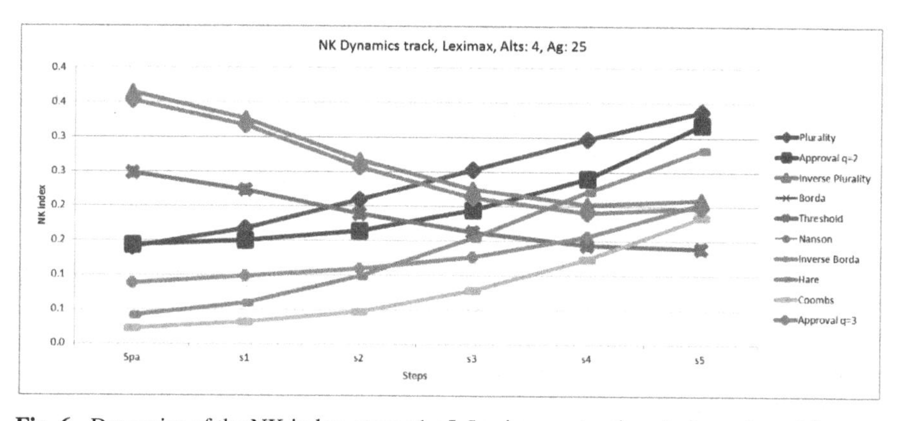

Fig. 6. Dynamics of the NK index, steps s1-s5, Leximax extension, 4 alternatives, 25 agents

For 5 alternatives, the degree of manipulability of the Coombs and Hare rules is also minimal at the initial stage of the Spa – s5 trajectory. However, for these rules, along with Inverse Borda, Nanson, Plurality, and Approval q = 2, manipulability increases, while for the Threshold, Approval q = 3, and Inverse Plurality procedures, the dynamics are more complex. The manipulability of these three procedures decreases from Spa to s4 but then increases towards s5 for some cases Fig. 7 and 8

It can be observed that the Threshold rule becomes least manipulated at point s5. Such dynamics are typical for a sufficiently large number of agents Fig. 8.

In the next Table 4 and Fig. 9, the results are presented in a slightly different manner. The current step on the dynamic track is fixed at s5, and the degree of manipulability in Table 4 is presented for several numbers of agents placed in columns. As one can see, for the Risk-averse extension, when the alternatives are shifted to point s5, closest to the

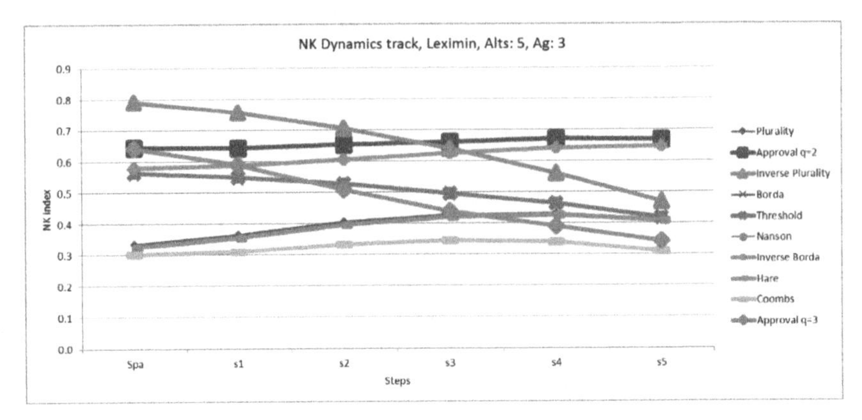

Fig. 7. Dynamics of the NK index, steps s1–s5, PWorst extension, 5 alternatives, 3 agents

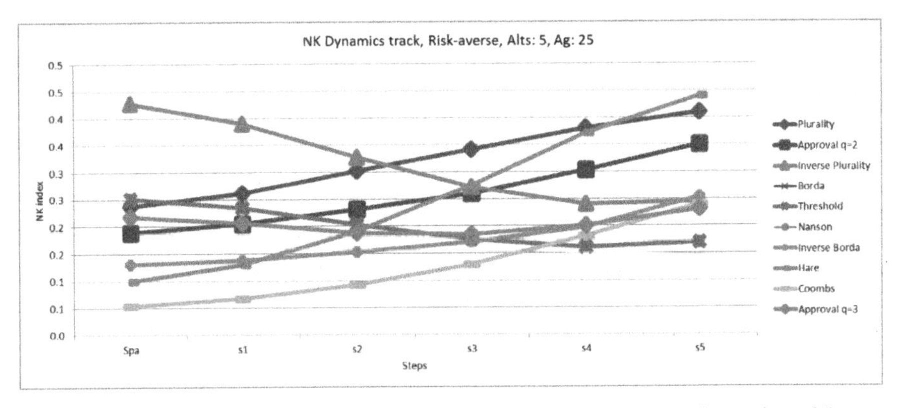

Fig. 8. Dynamics of the NK index, steps s1-s5, Risk-averse extension, 5 alternatives, 25 agents

midpoint of the agents, the Threshold rule is minimally manipulated as the number of agents increases beyond 7. For a smaller number of agents in this case, the Coombs rule is the least manipulable.

The result for 4 alternatives, presented in Fig. 10, shows slight differences from that for 5 alternatives, but the main tendency is the same.

6 Discussion and Future Considerations

In analyzing the manipulability dynamics of positional decision-making procedures within the dynamic voting model based on the two-dimensional Downsian framework, distinct patterns emerge for cases with 3, 4, and 5 alternatives. These patterns highlight the impact of the number of agents on the manipulability of various aggregation procedures.

For the cases of 4 and 5 alternatives, there are two groups of the aggregation procedures. The first group consists of the three procedures, Inverse Plurality, Approval, q = 3, and the Threshold procedures. The manipulability of these procedures decreases as the number of agents increases. This trend holds consistently across all considered

Table 4. NK index, step s5, Risk-averse extension, 5 alternatives

	4	7	10	13	16	19	22	25
Plurality	0.54	0.65	0.60	0.58	0.52	0.48	0.42	0.41
Approval q = 2	0.58	0.65	0.60	0.52	0.48	0.42	0.39	0.35
Inverse Plurality	0.55	0.59	0.51	0.39	0.34	0.29	0.32	0.25
Borda	0.70	0.60	0.51	0.44	0.38	0.32	0.28	0.25
Threshold	0.45	0.41	0.34	0.28	0.23	0.20	0.20	0.17
Nanson	0.70	0.60	0.51	0.44	0.38	0.32	0.28	0.25
Inverse Borda	0.70	0.60	0.51	0.44	0.38	0.32	0.28	0.25
Hare	0.53	0.48	0.54	0.52	0.50	0.47	0.44	0.44
Coombs	0.44	0.41	0.42	0.35	0.34	0.29	0.27	0.24
Approval q = 3	0.50	0.45	0.40	0.34	0.31	0.28	0.26	0.23

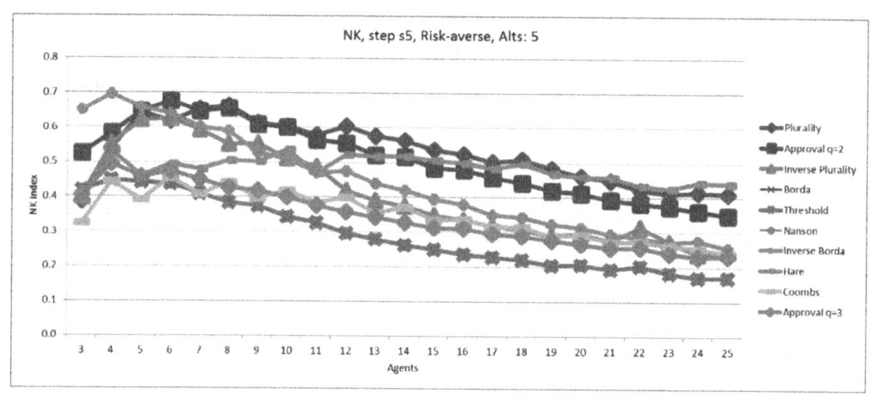

Fig. 9. NK index, step s5, Risk-averse extension, 5 alternatives, 3–25 agents

preference extensions, particularly becoming pronounced with larger numbers of agents (e.g., exceeding 10).

For the 3 alternatives the two groups of the procedures can be also observed. The only difference is that the manipulability of Inverse Plurality, Approval q = 3, and the Threshold procedures show a marginal decrease in manipulability, particularly for a small number of agents (less than 5). As the number of agents increases, the manipulability of these three procedures also shifts toward an increasing trend, aligning with the behavior observed for the other procedures in the group.

These observed manipulability patterns shed light on the intricate dynamics of positional decision-making procedures within the context of the dynamic voting model. The interplay between the number of alternatives, agents, and specific aggregation rules contributes to the nuanced landscape of manipulability. Further research and exploration

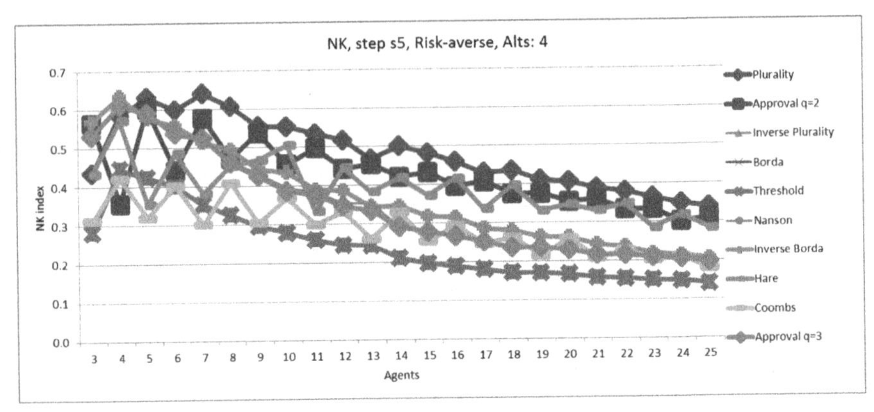

Fig. 10. NK index, step s5, Risk-averse extension, 4 alternatives, 3–25 agents

may delve into the underlying mechanisms driving these observed patterns and explore potential implications for real-world voting systems.

While our study primarily focused on individual manipulation, future research avenues may explore coalitional manipulability models, introducing collaborative strategies among agents. Investigating such models could offer additional insights into the collective dynamics of manipulability within the Downsian framework.

In conclusion, the manipulability dynamics observed in our study provide valuable insights into the behavior of positional decision-making procedures. These findings contribute to the broader understanding of strategic interactions in voting systems, emphasizing the need for comprehensive analyses to inform the design and evaluation of robust decision-making processes.

7 Conclusion

The general observations obtained from the results of the evaluation of the degree of manipulability for positional decision-making procedures in a dynamic voting model in the two-criteria Downsian model for 3, 4, and 5 alternatives and for the number of agents from 3 to 25 can be presented as follows.

For the 3 alternatives, the Plurality rule, Inverse Borda, and Nanson show an increase in manipulability when moving from the starting point Spa to point s5. The least manipulable at the Spa point is the Hare rule. The low degree of manipulability of this rule is preserved along the path of the movement to s5. For the Threshold and Coombs procedures, there is a noticeable slowdown in the growth of the degree of manipulability. At the s5 point, the degrees of manipulability of these two rules become close.

For 4 alternatives, with a sufficiently large number of agents, the Coombs and Hare rules are the least manipulable ones in the initial section of the trajectory. These procedures, along with Inverse Borda, Nanson, Plurality, and Approval $q = 2$ rules, show an increase in manipulability as they move towards s5. On the other hand, the rules Threshold, Approval $q = 3$, and Inverse Plurality have lower manipulability. The manipulability of the Threshold rule at point s5 is less than the manipulability of the Coombs rule. The manipulability of the Approval q = 3 and Inverse Plurality rules is greater than that of the Coombs rule in s5 and less than the manipulability of the Hare rule.

For 5 alternatives, the degree of manipulation of the Coombs and Hare rules is also minimal at the starting point Spa. For the Coombs, Hare, Inverse Borda, Nanson, Plurality, and Approval q = 2 rules, the degree of manipulability increases; however, for the Threshold, Approval q = 3, and Inverse Plurality rules, the dynamics are more complex.

Acknowledgements. This work was implemented in the framework of the Basic Research Program of the HSE University in 2024.

References

1. Novikov, S.G.: One dynamic problem in voting theory. I. Autom. Remote Control **46**(8), 1016–1026 (1985)
2. Novikov, S.G.: One dynamic problem in voting theory. II. Autom. Remote Control **46**(9), 1168–1177 (1985)
3. Downs, A.: An Economic Theory of Democracy. Harper, New York (1957)
4. Aleskerov, F., Kurbanov, E.: A Degree of Manipulability of Known Social Choice Procedures. In: Alkan, A., Aliprantis, C., Yannelis, N. (eds.) Current Trends in Economics: Theory and Applications, pp. 13–27. Springer, Heidelberg (1999)
5. Aleskerov, F.T., Karabekyan, D.C., Sanver, R., Yakuba, V.: Evaluation of the degree of manipulability of known aggregation procedures under multiple choice. J. New Econ. Assoc. **1–2**, 37–61 (2009). (in Russian)
6. Aleskerov, F.T., Karabekyan, D., Ivanov, A., Yakuba, V.: Individual manipulability of majoritarian rules for one-dimensional preferences. Procedia Computer Science (139). In: 6th International Conference on Information Technology and Quantitative Management. Elsevier, pp. 212–220 (2018). https://doi.org/10.1016/j.procs.2018.10.252
7. Aleskerov, F., Ivanov, A., Karabekyan, D., Yakuba, V.: Manipulability of majority relation-based collective decision rules. In: Czarnowski, I., Howlett, R.J., Jain, L.C. (eds.) IDT 2017. SIST, vol. 72, pp. 82–91. Springer, Cham (2018). https://doi.org/10.1007/978-3-319-59421-7_8
8. Karabekyan, D., Yakuba, V.: Manipulability of majoritarian procedures in two-dimensional downsian model. In: Morais, D., Fang, L., Horita, M. (eds.) Group Decision and Negotiation: A Multidisciplinary Perspective. GDN 2020. Lecture Notes in Business Information Processing, vol. 388, pp. 120–132 (2020). https://doi.org/10.1007/978-3-030-48641-9_9
9. McKelvey, R.D.: Intransitivities in multidimensional voting models and some implications for agenda control. J. Econ. Theory **12**(3), 474–482 (1976)
10. McKelvey, R.D.: General conditions for global in transitivities in formal voting models. Econometrica **47**(5), 1085–1112 (1979)
11. Kelly, J.: Almost all social choice rules are highly manipulable, but few aren't. Soc. Choice Welfare **10**, 61–175 (1993)
12. Nitzan, S. The vulnerability of point-voting schemes to preference variation and strategic manipulation. Public Choice 47, 349–370 (1985)

Collaborative and Responsible Negotiation Support Systems and Studies

Negotiation Platform for Supporting Multi-issue Bilateral Negotiations: The Case of an Offshore Wind Energy Company and a Fishing Community in the Northeast of Brazil

Eduarda Asfora Frej(✉), Gabriela Silva da Silva, Maria Luiza da Silva, and Danielle Costa Morais

Universidade Federal de Pernambuco–UFPE, CDSID-Center for Decision Systems and Information Development, Av. Acadêmico Hélio Ramos, s/n–Cidade Universitária, Recife 50.740-530, PE, Brazil
eafrej@cdsid.org.br

Abstract. This paper presents the development of an electronic negotiation platform to support multi-issue bilateral negotiation situations. The proposed platform has an embedded module for preferences elicitation based on the FITradeoff multicriteria method, to support negotiators with the evaluation of candidate packages when proposing offers or when deciding whether to accept or not an offer from its counterpart. The design and structure of the platform are described in detail in this paper. In order to illustrate the applicability of the proposed platform, a negotiation problem concerning the use of maritime areas in the Northeast region of Brazil is presented. The results of a negotiation between an Offshore Wind Turbine (OWT) company and a Fishing community is presented, in which conflicts with respect to four issues were dealt and a total of 256 packages were involved. After 7 negotiation rounds, an agreement was reached, and a post-settlement analysis was performed to verify the efficiency of the agreement package.

Keyword: Negotiation Support · Electronic Platform · Preferences Elicitation · Software architecture · FITradeoff method

1 Introduction

Multi-issue bilateral negotiations are characterized by situations in which two parties are involved in a negotiation problem which has more than one issue to be resolved. Multi-issue negotiations have a complex nature, since preferences of the negotiators with respect to those issues are usually conflicting [1]. Hence, eliciting negotiators preferences with regards to the issues involved is an important task in multi-issue negotiations, since it can help negotiators to evaluate the possible packages. In special, it can provide support for parties when proposing offers or when choosing to accept or not a counteroffer, which is what negotiators are truly seeking for [2].

M. Campos Ferreira et al. (Eds.): GDN 2024, LNBIP 509, pp. 117–129, 2024.
https://doi.org/10.1007/978-3-031-59373-4_10

In this context, negotiation platforms which have embedded support tools for preference elicitation outstands over common offer exchanging platforms, since they have the advantage of promoting such facilitation for the negotiation parties.

Multi-issue negotiations are analogous to multicriteria decision problems, in a sense that issues can be viewed as criteria of the problem, and the set of packages is the set of alternatives. In this sense, Multicriteria Decision Making/Aiding (MCDM/A) approaches can be applied to support the preferences elicitation of negotiators in multi-issue negotiations. Previous works have addressed the use of MCDM methods to aid preference elicitation in negotiation situations. The use of scoring systems with the direct rating approach is widely applied [3]; however, the establishment of scores may lead to some inconsistencies [4]. Several MCDM methods have been applied for negotiation support in the literature [5–7].

In order to avoid the use of scoring systems and to provide a flexible and less cognitively demanding elicitation process, Frej et al. [8] developed a negotiation protocol based on the FITradeoff multicriteria method [9–11] for preferences elicitation. This method works with partial information about users' preferences, without the need to stablish exact values for issues' weights, issues options or packages. Indirect information about issues' weights is obtained when negotiators answer questions put by the system considering tradeoffs amongst negotiation issues. A ranking of packages is computed according to the information about issues' weights, so that each party can analyze its own ranking when proposing offers or when choosing to accept or not a counteroffer. The model works based on a dynamic set of packages, in sense that, during the negotiation process, packages that are not interesting for both parties are eliminated from the process. Once the negotiators reach an agreement, a post-settlement analysis is conducted to verify the efficiency of the agreement solution [8, 12].

In this context, this paper aims to present the development of a web-based negotiation platform to operationalize the FITradeoff-based protocol, explaining the whole design and architecture of the tool. The aim of the platform developed in this paper is to support bilateral negotiations in the most varied contexts of problems. It aims to provide a favorable environment for conducting the elicitation of negotiators' preferences, with a decision support tool embedded into the platform.

In addition, this paper also presents the use of the developed platform applied to a negotiation process between an Offshore Wind Energy company and a Fishing community in the Northeast of Brazil, considering conflicts with respect to maritime area usage.

This paper is structured as follows. Section 2 presents the structure of the web-based negotiation platform. Section 3 presents the FITradeoff-based negotiation process, as well an explanation on how the negotiation platform works. Section 4 presents an application of use of the platform with a case of an Offshore Wind Energy company and a Fishing community in the Northeast of Brazil. Finally, final remarks are made in Sect. 5.

2 Web-Based Negotiation Platform

In this work, a web-based platform was constructed to operationalize the negotiation protocol developed by Frej et al. [8], in which bilateral multi-issue negotiations are carried out with preferences elicitation conducted with the FITradeoff method, considering a dynamic set of packages.

In general, software architecture represents the structure or set of structures which comprises the elements that will compose the software, their externally visible properties and their relationships [13]. The architecture of a software begins to be built in the initial stages of a software development process, aiming to define and visualize the computational solution that will be implemented. This artifact is known as initial architecture and belongs to the scope of the problem. Its main characteristic is to describe the solution at a high level of abstraction and it is used by various stakeholders as a basis for decision making [14].

The negotiation platform presented in this paper was developed in Delphi environment, using oriented Pascal based computational language. For the relational database management system, MySQL was used so that it was possible to provide the necessary tools to store and retrieve negotiations data using Structured Query Language (SQL). Communication data obtained from possible messages exchanged by the negotiation parties was also stored in this database. The platform is available for access on the most varied types of operating systems and devices, as long as it has the availability of an Internet connection and a web browser. The FITradeoff for Negotiation platform is directed to support bilateral electronic negotiations, and can be used to solve problems in different contexts using preference modeling with partial information. The FITradeoff multicriteria method was embedded into the platform to provide negotiation support, according to the protocol developed by [8].

In the negotiation platform, users (negotiators and mediator) are presented with user interfaces that allow them to interact with the system. Therefore, the platform is structured in three main phases: I - initialization, II - application, and III - data storage, and these incorporate the software elements, the connectors and the organization/configuration of the system.

In the initialization phase (I) there are the initial stages of entry into the platform, where the interfaces are presented so that user registrations are carried out, which can be either negotiators or mediators. After registration, validation and login to the system is performed, a new negotiation can be registered and started. All negotiations in which the user is a part of are stored in a list on this home screen, with all the data, information, exchange of offers and messages, so that the user can access whenever he/she needs it. The application phase (II) is where the negotiation actually takes place, all the procedures and methods for preferences elicitation are implemented together with the resources available at this stage so that the negotiations are made possible. Finally, the third phase consists of storing data (III) of the entire negotiation. All user and related information, trading style evaluations, data referring to negotiators' preferences and all generated numerical information, data on exchange of offers and counteroffers, message exchange data, data regarding the procedure of optimization and various other information associated with each of the activities performed during a negotiation within the platform.

As mentioned before, the relational database management used is MySQL so that it is possible to provide the necessary tools to store and retrieve negotiation data using Structured Query Language (SQL), so the platform has an online server that stores all this information to ensure that there is no loss of data and that negotiations can be carried out via the web. It is important to highlight that negotiations can be carried out synchronously (with both parties connected simultaneously) or asynchronously (with the flexibility of the parties to carry out their activities on the platform whenever they wish). It is worth noting that as this is a platform developed for use on the web, all data processing is supported by the use of the internet.

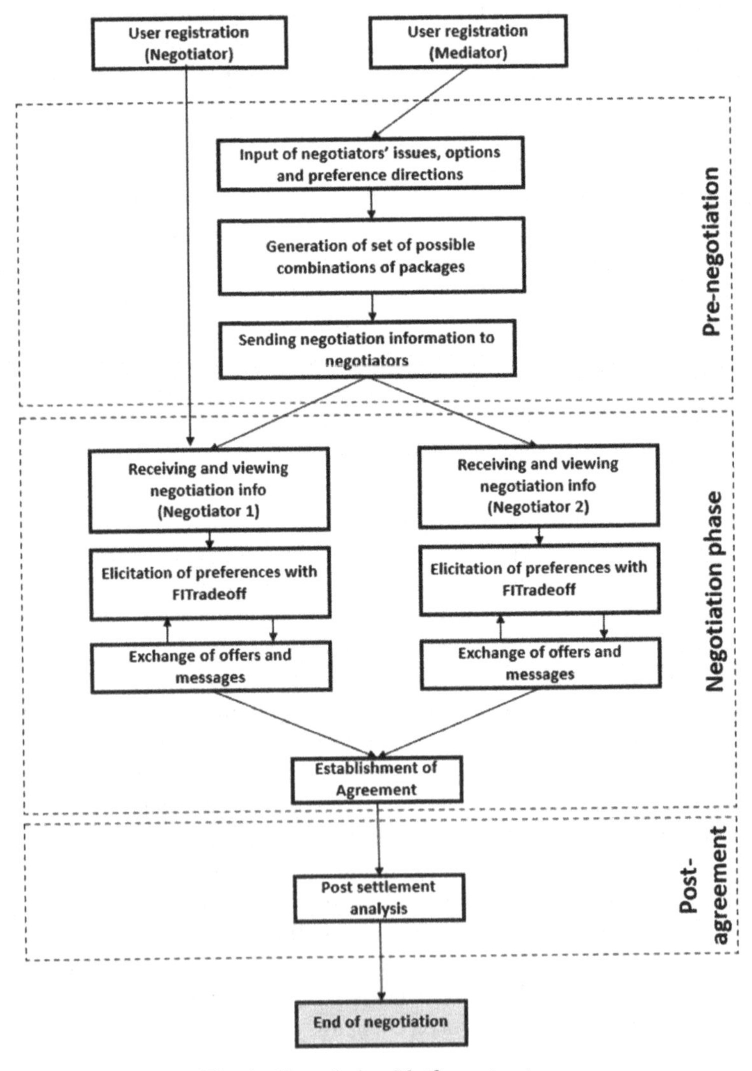

Fig. 1. Negotiation Platform structure

Finally, given the architecture presented above and aiming at a better understanding, a flowchart is presented below in Fig. 1, containing the flow of functioning and information of FITradeoff for negotiation and its respective phases: the pre-negotiation phase, the negotiation phase and the post-agreement phase.

It should be highlighted that arrows between the elicitation of preferences stage and the offer exchanging are bidirected. This means that the negotiators conduct the preferences elicitation process during the negotiation phase, with the possibility of answering more preference questions in order to have a more refined ranking of packages, as presented by Frej et al. [8]. The possibility of providing additional preference information during the negotiation process (i.e., in parallel with the offer exchanging process) gives to the negotiators more flexibility, in a sense that they need to provide preference information only until the point that the ranking obtained is sufficient for him to propose an offer and/or to decide whether to accept or not a counteroffer [8].

The detailing of each phase displayed in Fig. 1 will be further presented in the next section.

3 FITradeoff-Based Negotiation Process

3.1 Negotiation Support with FITradeoff

A negotiation protocol with preferences elicitation conducted based on the FITradeoff multicriteria method was developed by Frej et al. [8], in order to facilitate the elicitation process for negotiators, in sense that the establishment of exact scores are not needed. Instead, partial information about users' preferences is obtained through an interactive question-answering process, in which the user answers strict preference questions considering tradeoffs between issues' options.

In the FITradeoff-based protocol, the preferences elicitation is conducted in an interactive manner, in which each negotiator answers preference questions considering tradeoffs between the issues involved. The questions put for the user consists of a comparison between two hypothetical packages, with different values for issues options. In general, the first hypothetical package presents an intermediate outcome on a higher-ranked issue, and the worst outcome for all other issues. The second hypothetical package presents the best possible outcome for a lower-ranked issue and worst outcome for all other issues. The user should state which of these hypothetical packages he/she prefers. This information is then converted into an inequality that relates the values of the weights of the issues being evaluated. The inequalities obtained from a space of weights, which is the set of all possible weights vectors for issues' weights, compatible with the user's preferences [8].

These inequalities enter as constraints for linear programming models, that run searching for dominance relationships between candidate packages. Those dominance relations are derived into a ranking of packages, which can be used by the negotiators to aid the proposition of offers or to aid the decision on whether to accept or not an offer from the counterpart. Negotiators exchange offers based on their ranking of packages, and refinements on the ranking are conducted as long as they provide additional preference information [8].

The proposed model works based on a dynamic set of packages. Packages that are not interest for both parties are eliminated from the process. A package is eliminated from the negotiation if i) a package is offered by a party and refused by the counterparty; or ii) a package is dominated by a previously refused offer, in both parties' rankings. The elimination of irrelevant packages during the negotiation process can take negotiators to a convergence on a agreement package in a more efficient way. In the end, once the negotiators reach an agreement, a post-settlement analysis is conducted in order to verify the efficiency of the agreement solution [8, 12].

The purpose of this work is to present a negotiation platform to operationalize the FITradeoff-based protocol for negotiation support. In the literature of negotiation analysis, several other negotiation platforms have been developed [15]. The INSPIRE system [3] is a well-known negotiation platform which works based on the construction of a scoring system in which the negotiators should stablish values for issues, issues options and complete packages in order to build a utility function to evaluate the packages. The Negoisst system [16] is a negotiation platform devoted to B2B negotiation support, oriented for both documentation and communication purposes. The more recently developed e-nego system [17] works based on holistic evaluations for preferences elicitation, instead of using scores, using the MARS and UTASTAR approaches.

In this context, the idea of this work is to present a new negotiation platform to operationalize the negotiation-based protocol based on the FITradeoff multicriteria method for preferences elicitation, developed by [8]. The contribution of this work relies on providing to users the possibility of taking advantage of all the features of the FITradeoff method and its support to conduct negotiations. The FITradeoff method is already widely used in several application contexts for multicriteria decision-making situations [18], considering different decision problematics: choice [9], ranking [11], sorting [19] and portfolio [20, 21]. Expanding the use of the method for multi-issue negotiation support and the development of a negotiation platform with the FITradeoff protocol embedded is therefore the main contributions of this work to the literature of negotiation support.

The following subsection explains how the platform for operating the FITradeoff-based protocol for negotiation support works, considering both mediator and negotiator tasks.

3.2 Negotiation Platform

This section aims to describe how the FITradeoff-based negotiation protocol is operationalized through a negotiation platform developed in this work. This explanation is conducted two-fold: i) from the perspective of the mediator role; ii) from the perspective of the negotiators' role.

Starting from the mediator perspective, the mediator participates in basically two phases in FITradeoff for Negotiation: the pre-negotiation phase and the post-settlement phase. Between these two phases (i.e., during the actual negotiation phase), the mediator follows what is happening, but it does not actively participate in the process.

In the pre-negotiation phase, which takes place outside the platform, it is necessary to establish; i) negotiation parties; ii) negotiation issues; iii) negotiation issues' options. Those three elements are inputs for the platform. Once the parties are defined, the issues to be addressed during the negotiation process and which set of values each of these issues can admit are also determined. Furthermore, without the other party knowing, it is necessary to define the directions of preference for each party to the negotiation, that is, whether the issues will be of maximization or minimization for each negotiator individually.

The mediator is responsible for creating the negotiation room, inputting the negotiation data and subsequently sending this information to negotiators 1 and 2. To create the negotiation room, the mediator must include the two negotiators (previously defined) who will compose the negotiation. After that, where all the data of the problem will be inputted, these data refer to the issues that will be addressed (issues can be measured on either a natural or constructed scales), the options in each of issues, and the individual preferences directions for each negotiator (maximizing or minimizing). Then, the initial set of all possible packages is generated based on the combinations of issues options; these being the packages available to be negotiated. After completing these steps, the information can be sent to the negotiators. Hence, an e-mail is sent to the negotiators indicating the creation of a negotiation room and presenting initial instructions. All these phases are processed via internet and their respective data are stored in the server's database.

After completing these initial steps, the actual negotiation will take place. During the negotiation itself, the mediator role is only monitoring the negotiation process, without participating or interfering with it, at this time, it has access to a mediation Dashboard that contains information related to the development of the negotiation, this information is updated as negotiators go forward on the offer exchanging process.

The negotiation phase consists of the offer exchanging process, during which the negotiators can individually elicit preferences based on the FITradeoff protocol. This process is summarized in Fig. 2.

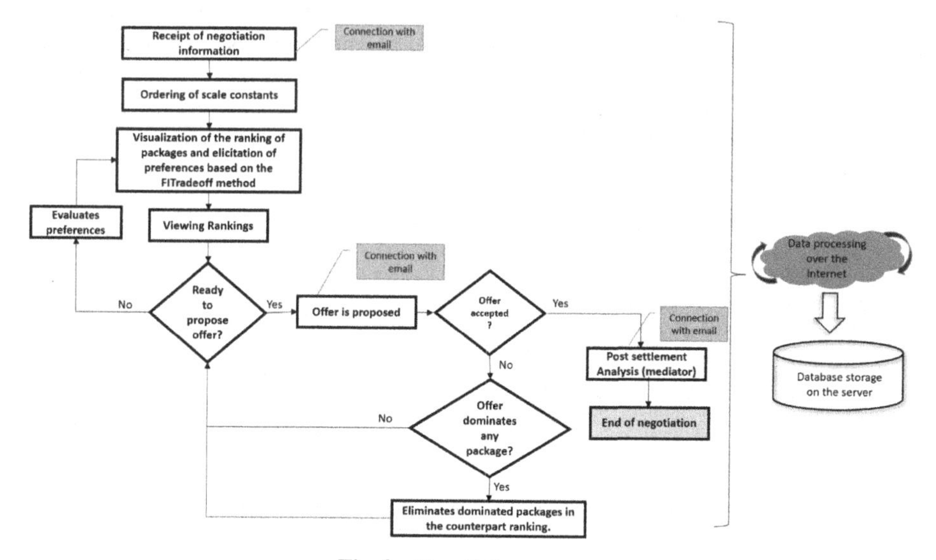

Fig. 2. Negotiation process

The process starts when the negotiator receives information regarding the negotiation problem, which were entered into the system by the mediator, so that they can start the process. The first part consists of carrying out the elicitation of preferences following the FITradeoff method for the ranking problematic [11]. Whenever a negotiator is already ready to propose an offer, the process of exchanging offers and counteroffers can be started. As the offer exchanging process goes on, whenever an offer is refused by the counterpart, the elimination of dominated packages is carried out, following the negotiation protocol proposed by Frej. et al. [8]. It is also available for the negotiators the possibility of exchanging informal messages, if they wish. This process goes on until an agreement is achieved.

When the negotiators reach an agreement, the post-settlement phase is carried out. At this stage, the mediator verifies if the agreement package is a pareto-optimal solution of the problem. Since the mediator have access to both parties rankings of packages – obtained in the preferences elicitation process with FITradeoff -, a verification on whether there is any package that is better than the agreement packages, in both rankings. If no better package can be found, the agreement package is a pareto-optimal solution for the problem; otherwise, the agreement package can be considered dominated, and better packages considering both negotiators' rankings can be reached. In this sense, when better solutions can be achieved, these solutions are displayed for the negotiation parties, via e-mail communication, so that they can improve the solution obtained, if so they desire. It should be highlighted that negotiators are free to accept or not the suggested solution given by the mediator. On the other hand, when no better package can be found, feedback is also provided for negotiators, informing them that their solution is indeed a Pareto-Optimal solution.

In all negotiation phases (room creation, exchange of offers and messages, negotiation completion and post-settlement), users are always notified by e-mail. All these phases are processed via the internet and their respective data are stored in the server's database. Thus, based on the information presented above, there is the establishment of FITradeoff for Negotiation platform. Figure 3, below, shows the resources incorporated into the platform for its full operation.

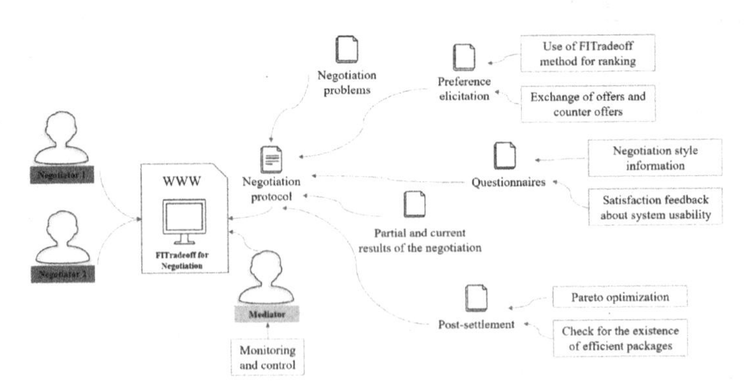

Fig. 3. FITradeoff for Negotiation Platform

4 Application: Negotiation Between an Offshore Wind Energy Company and a Fishing Community in the Northeast of Brazil

The negotiation between offshore wind turbine (OWT) companies and fishermen evolve several conflicts related to maritime area usage. In this context, the FITradeoff for negotiation platform was applied to facilitate constructive dialogue between an offshore wind energy company and a fishing community in the Northeast region of Brazil. Representatives from the OWT company and fishing community structured their preferences and negotiations using the FITradeoff method.

The following issues were considered in the negotiation process:

Issue 1. Financial Compensation (natural scale, in Brazilian reais): The corporation should provide financial reparation to the fishing community for adverse effects, with amounts designated at four levels: $50,000; $80,000, $110,000; and $140,000. The corporation aims to minimize this compensation, in contrast to the fishermen who seek maximization.

Issue 2. Reduction of Turbine-Impacted Area (constructed scale, 4 levels):

i. Level 1: Minimal effort to reduce the impacted area, covering more than 30% of the traditional fishing area.
ii. Level 2: Moderate effort, aiming to impact no more than 30% of the traditional fishing area.
iii. Level 3: Substantial effort, limiting the impacted area to 15% of the traditional fishing area.
iv. Level 4: Maximum effort, with less than 10% of the traditional fishing area being affected by turbine installation.

Issue 3. Sustainable Fishing Development (constructed scale, 4 levels):

i. Level 1: Investment of up to $100,000 in technologies and training for sustainable fishing, focusing on traditional methods.
ii. Level 2: Investment of up to $400,000, introducing basic technologies and training in sustainable fishing practices.
iii. Level 3: Investment of up to $900,000, implementing advanced technologies and comprehensive training programs to promote sustainable fishing.
iv. Level 4: Investment of more than $900,000, including ongoing research, adoption of innovative technologies, and environmental education to ensure long-term sustainability of fishing in the region.

Issue 4. Marine Life Preservation and Impact Mitigation (constructed scale, 4 levels):

i. Level 1: Implementation of basic measures for marine life protection, such as limited exclusion zones and basic species monitoring, with an investment of up to $100,000.
ii. Level 2: Investment of up to $400,000 to implement additional measures, such as comprehensive monitoring programs and noise mitigation measures.
iii. Level 3: Commitment of up to $900,000, with the establishment of marine protected areas and investment in technologies to reduce the impact of turbines on marine life.

iv. Level 4: Commitment of more than $900,000, including the implementation of advanced mitigation strategies, such as early warning systems to prevent collisions with marine animals and habitat restoration programs.

It should be highlighted here that negotiations are conducted considering the complete set of packages, which is obtained based on the combination of all options of the issues, through a combinatorial approach. In this sense, we consider a discrete set of options for each issue (i.e., salient options only), since continuous variables would not enable us to generate all possible packages. The negotiation packages shown by the system for the parties are composed by the options of the issues previously defined in the problem structuring stage. However, the platform enables parties to exchange messages, in sense that they could communicate and possibly set an agreement value different from the previously defined options. The main goal of the system is to provide support for the parties, but of course they have freedom to decide and adapt according to their own needs and wishes.

Considering all possible combinations of issues' options, a total of 256 packages are involved in this problem. Each negotiator had a preference direction for each issue, with the company preferring to minimize (as all issues imply costs for the company) and the fishermen preferring maximization, showing the need to balance economic, environmental, and community interests.

The FITradeoff method was used to model the negotiators' preferences based on comparisons between hypothetical packages, following the protocol developed by Frej et al. [8], thereby finding a preference ranking for both parties. Negotiators could observe their preference rankings during the negotiation stage and make offers.

Table 1 shows that fishermen and the wind turbine company made alternating offers during negotiation. The outcomes of each offer were measured by various factors, including the number of eliminated packages (nEPo), the cumulative number of eliminated packages (CnEP), the percentage of elimination (nEP%), the number of remaining packages (NRP) and the percentage of remaining packages (NRP%). Additionally, the number of elicitation questions answered by each negotiation in the elicitation process with FITradeoff is also displayed; NQN1 refers to the number of questions answered by the Offshore Wind Turbine (OWT) company, and NQN2 pertains to the number of questions answered by the fisherman.

The analysis of Table 1 reveals the progression of the negotiation. There is a noticeable increase in both the number of packages eliminated and the percentage of elimination over time, suggesting a more refined filtering of options to those that best meet negotiators' preferences. A decrease in the number of remaining packages indicates a shift towards consensus. Furthermore, the consistent number of questions answered by both parties indicates balanced and active participation in negotiations.

An agreement was reached when the fishermen accepted package 155, which was last offer from the wind turbine company after analyzing its position and associated values. Package 155 has the following values for issues options: Financial Compensation (R$ 110.000) Reduction of Turbine-Impacted Area (2) Sustainable Fishing Development (3) Marine Life Preservation and Impact Mitigation (3). After reaching an agreement, a post-settlement analysis is performed by the mediator, in order to verify whether there is or not a better solution for both parties. In this analysis, the mediator verifies the packages

Table 1. Summary of negotiation

Offer	Negotiator	Package	nEPo	CnEP	nEP%	NRP	NRP%	NQN1	NQN2
1°	OWT	Package1	1	1	0%	255	100%	5	8
2°	Fisherman	Package256	1	2	1%	254	99%	5	8
3°	OWT	Package10	20	22	9%	234	91%	5	8
4°	Fisherman	Package252	1	23	9%	233	91%	5	8
5°	OWT	Package74	32	55	21%	201	79%	5	8
6°	Fisherman	Package220	40	95	37%	161	63%	5	8
7°	OWT	Package155	Accepted package						

rankings obtained for both negotiators with the FITradeoff method, and it is verified if there is any solution that dominates the agreement package (in this case, package 155) in both rankings. In this case, no better package was found in the post settlement phase and the negotiation has ended here.

5 Final Remarks

This paper presented the development of a web-based negotiation platform to multi-issue bilateral negotiation situations. The preferences elicitation protocol was implemented in the platform following the ideas of Frej et al. [8], with the FITradeoff method being incorporated as an embedded module for the negotiation support tool. The whole design and structure of the platform was detailly presented throughout the negotiator and mediator operation tasks.

The use of the platform was illustrated with a negotiation problem between an Offshore Wind Turbine (OWT) company and a Fishing community in the Northeast of Brazil, in which conflicts with respect to maritime area usage were involved. The results of the application have shown that an agreement was reached after 7 negotiation rounds (i.e., a total of 7 offers were made until the fisherman accepts the offer made by OWT company). Regarding the preferences elicitation process, 5 FITradeoff questions were answered by the OWT company, and 8 FITradeoff questions were answered by the fisherman. This number consist of the number of preference questions answered by each of them in order to obtain a ranking of the packages, according to the algorithm proposed by [8].

Since the proposed approach works based on a dynamic set of packages, such that packages not interest for both parties are eliminated from the process, a certain number of packages was eliminated during the negotiation rounds. At the end of the process, 37% of the packages had been eliminated. This approach has shown to be useful for handling this negotiation situation, and it also has potential for applications in order negotiation contexts, as long as multiple issues are involved.

For future work, simulation studies should be conducted in order to analyze the speed of convergence of the approach considering the elimination of packages during the process. In addition, behavioral studies with neuroscience tools could also be conducted to analyze behavioral issues about the negotiators when using this approach. The present paper aimed to present the design of the platform and an illustrative use of it; hence, future studies should explore the validation of the suitability of the platform for improving negotiation support, with multiple cases studies to analyze aspects such as usability and performance.

Finally, it is important to state that the platform presented in this paper would benefit from some improvements and adjustments, that should be carried out in the future. For instance, the definition of issues and issues options are currently made outside of the platform. It would be interesting to consider the incorporation of a pre-negotiation module into the platform, in order to enable these elements to be defined via web in a collaborative manner by the parties.

Acknowledgments. The authors are most grateful for CNPq, FACEPE and CAPES, for the financial support provided.

References

1. Raiffa, H., Richardson, J., Metcalfe, D.: Negotiation analysis: the science and art of collaborative decision making. Harvard University Press (2002)
2. Vetschera, R., Filzmoser, M., Mitterhofer, R.: An analytical approach to offer generation in concession-based negotiation processes. Group Decis. Negot. **23**(1), 71–99 (2014)
3. Kersten, G.E., Noronha, S.J.: WWW-based negotiation support: design, implementation, and use. Decis. Support Syst. **25**(2), 135–154 (1999)
4. Kersten, G.E., Roszkowska, E., Wachowicz, T.: How Well Agents Represent their Principals' Preferences: The Effect of Information Processing, Value Orientation, and Goals (2020)
5. Wachowicz, T., Błaszczyk, P.: TOPSIS based approach to scoring negotiating offers in negotiation support systems. Group Decis. Negot. **22**(6), 1021–1050 (2013)
6. Wachowicz, T.: Decision support in software supported negotiations. J. Bus. Econ. Manag. **4**, 576–597 (2010)
7. Silva Filho, J.L., Costa Morais, D.: Negotiation protocol based on ordered weighted averaging and Fuzzy metrics. J. Organ. Comput. Electron. Commer. **29**(3), 190–208 (2019)
8. Frej, E.A., Morais, D.C., de Almeida, A.T.: Negotiation support through interactive dominance relationship specification. Group Decision Negotiation, 1–30 (2022)
9. De Almeida, A.T., Almeida, J.A., Costa, A.P.C.S., Almeida-Filho, A.T.: A new method for elicitation of criteria weights in additive models: flexible and interactive Tradeoff. Eur. J. Oper. Res. **250**(1), 179–191 (2016)
10. De Almeida, A.T., Frej, E.A., Roselli, L.R.P.: Combining holistic and decomposition paradigms in preference modeling with the flexibility of FITradeoff. CEJOR **29**(1), 7–47 (2021)
11. Frej, E.A., de Almeida, A.T., Costa, A.P.C.S.: Using data visualization for ranking alternatives with partial information and interactive tradeoff elicitation. Oper. Res. Int. J. (2019). https://doi.org/10.1007/s12351-018-00444-2
12. Correia, L.M.A., Frej, E.A., Ribeiro, M.L.S., Morais, D.C.: Eliciting Preferences with Partial Information in Multi-issue Negotiations: An Analysis of the FITradeoff-Based Negotiation Protocol. In: International Conference on Group Decision and Negotiation, pp. 17–30. Springer International Publishing, Cham (2022)

13. Bass, L., Clements, P., Kazman, R.: Software Architecture in Practice, Second Edition, Addison Wesley (2003)
14. Clements, P.,et al.: Documenting Software Architectures. Addison Wesley (2004)
15. Kersten, G.E., Lai, H.: Negotiation support and e-negotiation systems: An overview. Group Decis. Negot. **16**, 553–586 (2007)
16. Schoop, M., Jertila, A., List, T.: Negoisst: a negotiation support system for electronic business-to-business negotiations in e-commerce. Data Knowl. Eng. **47**(3), 371–401 (2003)
17. Wachowicz, T., Roszkowska, E.: Holistic preferences and prenegotiation preparation. Handbook of Group Decision and Negotiation, pp. 255–289 (2021)
18. De Almeida, A.T., Frej, E.A., Roselli, L.R.P., Costa, A.P.C.S.: A summary on fitradeoff method with methodological and practical developments and future perspectives. Pesquisa Operacional **43**, e268356 (2023)
19. Kang, T.H.A., Frej, E.A., de Almeida, A.T.: Flexible and interactive tradeoff elicitation for multicriteria sorting problems. Asia-Pacific J. Oper. Res. **37**(05), 2050020 (2020)
20. Frej, E.A., Ekel, P., de Almeida, A.T.: A benefit-to-cost ratio based approach for portfolio selection under multiple criteria with incomplete preference information. Inf. Sci. **545**, 487–498 (2021)
21. Marques, A.C., Frej, E.A., de Almeida, A.T.: Multicriteria decision support for project portfolio selection with the FITradeoff method. Omega **111**, 102661 (2022)

Allocation of Recurring Fixed Costs According to Partners' Varying Revenues in Professional Services

Eugene Khmelnitsky and Yigal Gerchak(✉)

Department of Industrial Engineering, Tel-Aviv University, 69978 Tel-Aviv, Israel
{xmel,yigal}@tauex.tau.ac.il

Abstract. We consider professional service providers, such as lawyers and physicians, who essentially work independently, but share an office or clinic. They need to jointly cover recurring fixed costs, like rent and staff wages, in every period. It is often argued that such costs should be allocated based on usage of the facilities. A more quantifiable indirect indicator of usage is the revenues. As partners' revenues fluctuate over time, it is plausible to require, as we do in this study, that their shares in the fixed costs fluctuate accordingly. Had partners' revenues not fluctuated, the deterministic cost allocation problem would be mathematically equivalent to the bankruptcy problem. Ours is a dynamic stochastic model of cost allocation for which we examine some ex-ante mechanisms based on partners' portions of total revenue. Monotonicity, proportionality, and other properties of the suggested mechanisms are studied.

Keywords: professional services · cost allocation · facility sharing · revenue ratios · gamma distributions

1 Introduction

Several cooperating entities whose revenues and usage of facilities vary over time may need to cover recurring fixed costs in each period. One such scenario are service providers (e.g., physicians, lawyers of similar or different specialties), who essentially work independently (have their own patients/clients). Professional services like these are an understudied industry. Young [1] argued that "When two doctors share an office, for example, they need to divide the cost of office space, medical equipment, and secretarial help". They thus need to jointly cover this recurring cost in every period (see also Young [14]). A rather "natural" way to allocate the costs is by usage of the facilities which give rise to these costs. That is done, for instance, in determining aircraft landing fees at airports [2, 3]. Such method is not very practical in our setting, as monitoring secretary's or nurse's service activities for each partner is functionally difficult and time consuming. Rather, one can use instead the partners' revenues, which reflect their usage of the facilities, and that is indeed what we use.

More generally, the problem of fixed costs allocation arises in real-life situations where multiple decision makers ("partners") create a common platform, which supports

M. Campos Ferreira et al. (Eds.): GDN 2024, LNBIP 509, pp. 130–141, 2024.
https://doi.org/10.1007/978-3-031-59373-4_11

their production and/or service processes, and its costs need to be covered. A partner's share in the fixed costs associated with the maintenance of a platform typically includes a fixed charge, which is independent of the operational activity of the partners, and a charge which increases with the actual partner's usage of the platform in a given time period. In the scenario described above, the fixed charge is the office rent, in which case a partner will pay a fixed share of total office space. Unlike the fixed charge, the variable charge increases with the usage of the supporting personnel and office facilities in each period. This paper focuses on determining the variable charge and suggests using partners' fluctuating period's revenues as reflecting the extent of partner's operational activity. Had the partners' revenues not fluctuated, the deterministic version of the problem would be mathematically equivalent to the Bankruptcy Problem, also referred to as adjudication of conflicting claims, where an endowment is insufficient to meet all claims [4–6]. Ertemel and Kumar [7] and Xue [8] introduce uncertainty of claim amounts (see also Thomson [6], Sec.6.3). The focus is on balancing "waste" (unused endowment) and shortage. These are one-time scenarios, as opposed to our multi-period setting, and thus our requirement of time-average fairness (proportionality) is not relevant there.

We examine several mechanisms for the allocation of fixed costs that could be imposed, rather than having the partners negotiate one themselves. The mechanisms we examine are allocating costs according to ratios of a partner's expected or realized period's revenue to the sum of the partners' expected or realized revenues. We assume that efficiency of resource usage is fixed.

We seek mechanisms that satisfy monotonicity and fairness properties. Monotonicity requires allocating a higher share of the costs to a partner were her revenue to increase. Fairness is defined here as the requirement that the long-run average of a partner's share be equal to the ratio of her expected revenue to the sum of the partners' expected revenues ("proportionality"). The next sections present such plausible mechanisms and discuss their monotonicity and fairness properties. Specifically, we show that fairness is satisfied in particular cases when partners' revenues can be modeled as a fairly rich family of distributions of non-negative variables - Gamma. Generally, in order to make the mechanism fair, we suggest a new analytical method, which is based on a modification of the distribution function of a partner's revenue.

2 Literature Review

There are many sharing mechanisms or cost allocation methods proposed in the literature. Some are based on simple proportional rules and others are based on theoretical concepts from game theory [1, 4].

In determining aircraft landing fees at a particular airport, runway usage, being the source of wear and tear of the runway was proposed by Littlechild and Thompson [3] to be the guideline. The authors study these issues as applying to the Birmingham UK airport. They identified "the club principle" as a plausible philosophy of costs allocation: A two-part tariff which combines fixed costs with proportional costs due to runway usage. They make use of the Shapley value for such allocation.

Guajardoa and Ronnqvist [9] provide a comprehensive review of cost allocation methods used in collaborative transportation, including problems of planning, vehicle routing, traveling salesmen, distribution, and inventory.

3 Basic Axioms and Schemes

Let an n-partners partnership's fixed monthly cost be $C^{(n)}$. It is required that the fixed monthly costs be covered for each period regardless of random realizations of partners' revenues. By X_i and x_i we denote partner i's random revenue and its realization, $i = 1, \ldots, n$, and by $f_i(\cdot)$ and $F_i(\cdot)$ its pdf and cdf, respectively. $X_1, \ldots, X_n$ are assumed to be mutually independent, which is reasonable as partners' clients' groups do not overlap. We assume that $\sum_{i=1}^{n} E(X_i) > C^{(n)}$, so on the long run the fixed costs can be covered from revenues. However, this does not guarantee that the total revenue will suffice in every period. By $P_i^{(n)}$ and $p_i^{(n)}$ we denote partner i's random cost allocation and its realization, respectively.

It is desirable that the cost allocation mechanisms satisfy the following conditions (axioms) (cf. Billera and Heath, [10], Thomson [6]):

Non-negativity: partner's contribution is nonnegative, $p_i^{(n)} \geq 0\ \forall i$, at each period.

Full coverage: partners' contributions cover the fixed cost, $\sum_{i=1}^{n} p_i^{(n)} = C^{(n)}$, at each period.

Cost independence: partner's contribution grows linearly with $C^{(n)}$, i.e., the ratio $\frac{p_i^{(n)}}{C^{(n)}}$ is independent of the fixed cost.

Economies of scale: if $C^{(n)} \equiv C\ \forall n$, then $p_i^{(n)}$ is decreasing in n.

Monotonicity: partner's contribution, $p_i^{(n)}$, increases with her realized revenue, x_i, i.e., $\frac{\partial p_i^{(n)}}{\partial x_i} > 0$.

Proportionality: partners' mean contributions are proportional to their expected revenues,

$$E\left(P_i^{(n)}\right) = \frac{E(X_i)}{\sum_{j=1}^{n} E(X_j)} C^{(n)}.$$

Without requiring proportionality, i.e., if $E\left(P_i^{(n)}\right)$ is greater/smaller than $\frac{E(X_i)}{\sum_{j=1}^{n} E(X_j)} C^{(n)}$, then on the long run, partner i is expected to over/under contribute to costs[1].

We note that proportionality is consistent with monotonicity but is much more specific. We note also that the cost independence condition holds in all allocation schemes considered in this paper, and that the economies of scale condition is trivial when n is fixed.

3.1 Permanent Allocation Schemes

We shall now assume that the number of partners is fixed, so we shall suppress the superscript $(\boldsymbol{n})$ and normalize the fixed cost, $\boldsymbol{C}^{(n)}$, without loss of generality, to 1. A

[1] Requiring $E\left(P_i^{(n)}\right) = E\left(\frac{X_i}{\sum_{j=1}^{n} X_j}\right) C^{(n)}$ is also a possibility.

permanent allocation based on expected (long-run) revenues is thus:

$$p_i = \frac{E(X_i)}{\sum_{j=1}^n E(X_j)}. \tag{1}$$

Scheme (1) satisfies non-negativity, full coverage and proportionality. As revenue realizations are not captured by that scheme, monotonicity does not hold and that leads to scenarios where a partner whose random revenue is low several periods in a row will still need to pay the permanent high allocation; a high realized profit by another partner will not help her.

A permanent allocation based on the expected ratio of partners' revenues is:

$$p_i = E\left(\frac{X_i}{\sum_{j=1}^n X_j}\right). \tag{2}$$

Scheme (2) satisfies non-negativity, yet full coverage, monotonicity and proportionality may not hold unless that is used as the definition of proportionality.

3.2 Allocation Based on Realized Revenues

Consider a scheme where the fixed costs are allocated at each period according to the realized revenues of that period, i.e.,

$$p_i = \frac{x_i}{\sum_{j=1}^n x_j}. \tag{3}$$

Note that in this scheme a partner's contribution depends on the realized revenues of *all* partners.

Scheme (3) works as follows. Suppose that, in period t, $\sum_{j=1}^n x_j^t > 1$. Then partner i pays $\frac{x_i^t}{\sum_{j=1}^n x_j^t}$, and the amount that is left is.

$$x_i^t - \frac{x_i^t}{\sum_{j=1}^n x_j^t} = \frac{x_i^t\left(\sum_{j=1}^n x_j^t - 1\right)}{\sum_{j=1}^n x_j^t} > 0$$

If, in period t, $\sum_{j=1}^n x_j^t < 1$, then partner i pays x_i^t from revenue and completes the shortage $\frac{x_i^t}{\sum_{j=1}^n x_j^t} - x_i^t$ from its funds. Since we assumed that $\sum_{j=1}^n E(X_j) > 1$, mechanism (3) is feasible in expectation.

This scheme satisfies non-negativity, full coverage and monotonicity. Proportionality does not hold in general. However, in the special case where $X_i \sim Gamma(k_i, \theta, \cdot)\ \forall i$, with the pdf $f_i(x) = \frac{x^{k_i-1}e^{-x/\theta}}{\Gamma(k_i)\theta^{k_i}}$, $x \geq 0, k_i > 0, \theta > 0$, $E(X_i) = k_i\theta$, (see Fig. 1), ratio (3) satisfies proportionality, i.e.

$$E(P_i) = E\left(\frac{X_i}{\sum_{j=1}^n X_j}\right) \overset{(a)}{=} \frac{k_i}{\sum_{j=1}^n k_j} = \frac{k_i\theta}{\sum_{j=1}^n k_j\theta} \overset{(b)}{=} \frac{E(X_i)}{\sum_{j=1}^n E(X_j)},$$

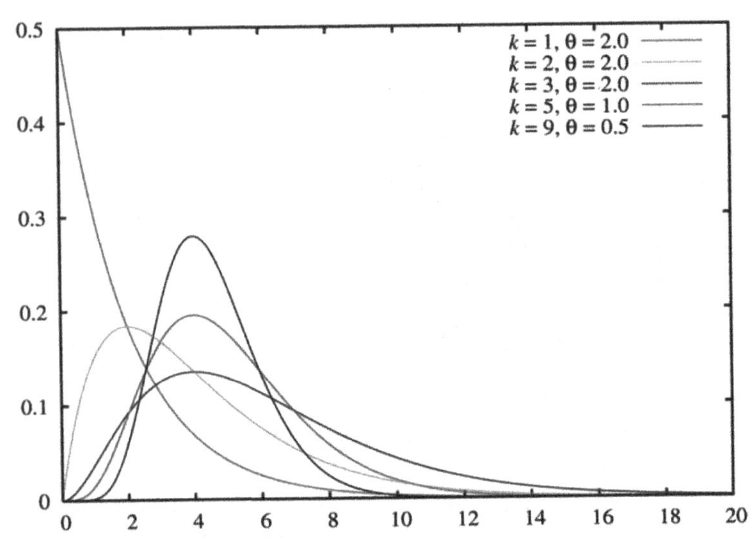

Fig. 1. Graphs of PDFs of Gamma distributions with various parameter values (reproduced from Wikipedia "Gamma distribution").

where equalities (a) and (b) follow from properties of the Gamma distribution [11]. We shall show how to transform any distribution to a Gamma one, and thus for that scheme to always satisfy proportionality.

A generalization of scheme (3) is.

$$p_i = \frac{x_i^\alpha}{\sum_{j=1}^n x_j^\alpha}, \alpha > 0.$$

Similar to (3), this scheme satisfies non-negativity, full coverage and monotonicity. In Sect. 4 we show a modification of that scheme causing proportionality to be satisfied.

Another scheme where the fixed cost is allocated according to the realized revenues is,

$$P_i = \frac{X_i}{\sum_{j=1}^n E(X_j)}. \tag{4}$$

Unlike scheme (3), here a partner's realized contribution depends only on its own realized revenue. Specifically, partner i, $i = 1, 2, \ldots, n-1$ pays

$$p_i = \frac{x_i}{\sum_{j=1}^n E(X_j)} \tag{5}$$

and partner $i = n$ completes that to 1, i.e., $p_n = 1 - \sum_{j=1}^{n-1} p_j$. Note that a risk-averse partner will prefer the (constant), expected value based (1), to the fluctuating realizations of (3) or (5).

Scheme (5) satisfies full coverage and proportionality conditions, while monotonicity holds only for partners i, $i \le n-1$. A problem with allocation (5) is that non-negativity does not hold, since $\sum_{j=1}^{n-1} p_j$ could exceed 1 in some time periods, in which case p_n becomes negative. Another problem is that partner n's revenue, x_n, has no impact, causing the allocation to be asymmetric (non-anonymous). Anonymity can be achieved by the partners alternating who is "last" in that scheme. Hence, the scheme can work when X_i, $i = 1, 2, \ldots, n$ are distributed in such a way that for any realization,

$$\frac{\left(\sum_{j=1}^{n} x_j\right) - x_i}{\sum_{j=1}^{n} E(X_j)} \leq 1, \ \forall i.$$

If this holds, the contribution of partner i, in the periods when she is last, is non-negative.

Example 1. Consider two partners whose incomes follow a Beta $(\alpha, 1)$ distribution, also called a power (α) distribution, $f_i(x) = \frac{\alpha(x-a_i)^{\alpha-1}}{(b_i-a_i)^{\alpha}}$, $i = 1, 2$, $0 \leq a_i \leq x \leq b_i$, $\alpha \geq 1$, $E(X_i) = \frac{a_i+\alpha b_i}{1+\alpha}$. Then no realization of $\frac{x_i}{E(X_1)+E(X_2)}$, $i = 1, 2$ exceeds 1 if $a_1 + a_2 \geq \max\{b_1 - \alpha b_2, b_2 - \alpha b_1\}$. ■

Example 2. Consider two partners whose income is distributed with $f_i(x) = \frac{2}{b_i - a_i} - \frac{2(x-a_i)}{(b_i-a_i)^2}$, $1 \leq a_i \leq x < b_i$, $i = 1, 2$, a triangular pdf decreasing in x. The expected revenue is $E(X_i) = \frac{2a_i+b_i}{3}$. In such a case the partners' shares are,

$$p_i = \frac{x_i}{E(X_1) + E(X_2)} \leq \frac{b_i}{\frac{2a_1+b_1}{3} + \frac{2a_2+b_2}{3}} = \frac{3b_i}{2a_1 + 2a_2 + b_1 + b_2},$$

which should not exceed 1. That is satisfied for both $i = 1$ and $i = 2$ if.

$$0 < b_1 - a_1 \leq a_2 + \frac{b_2}{2} \textit{ and } 0 < b_2 - a_2 \leq a_1 + \frac{b_1}{2}.$$

If $b_1 > b_2$ then need $a_1 + a_2 \geq b_1 - \frac{b_2}{2}$. Otherwise, need $a_1 + a_2 \geq b_2 - \frac{b_1}{2}$. ■

4 Modifications to Basic Schemes

This section suggests modifications to schemes (3) and (5), which satisfy all conditions. As noted, a major problem with (3) is that, generally, it does not satisfy proportionality.

Figure 2 illustrates the lack of fairness (lack of proportionality) of scheme (3) defined as the maximum relative difference between $\boldsymbol{E}\left(\frac{X_i}{\sum_{j=1}^{n} X_j}\right)$ and $\frac{E(X_i)}{\sum_{j=1}^{n} E(X_j)}$,

$$z = \max_i z_i, \text{ Where } z_i = \left(\frac{\boldsymbol{E}\left(\frac{X_i}{\sum_{j=1}^{n} X_j}\right)}{\frac{E(X_i)}{\sum_{j=1}^{n} E(X_j)}} - 1\right). \tag{6}$$

The plot of z is calculated approximately for $\boldsymbol{n} = 2$ and $\boldsymbol{X}_1$ and $\boldsymbol{X}_2$ both distributed *LogNormal*.

An analytical approximation of z_i can be obtained by accounting for the first two terms of the Taylor expansion of the expectation of a reciprocal of a random variable and by the fact that the ratio of two independent random variables distributed *LogNormal* is also *LogNormal*. Let $\boldsymbol{c}$ denote the coefficient of variation, assumed for simplicity to be identical for $\boldsymbol{X}_1$ and $\boldsymbol{X}_2$. Then,

$$\boldsymbol{E}\left(\frac{\boldsymbol{X}_1}{\boldsymbol{X}_1+\boldsymbol{X}_2}\right)=\boldsymbol{E}\left(\frac{1}{1+\boldsymbol{X}_2/\boldsymbol{X}_1}\right)\approx\frac{1}{1+\boldsymbol{E}\left(\frac{\boldsymbol{X}_2}{\boldsymbol{X}_1}\right)}+\frac{\boldsymbol{Var}\left(\frac{\boldsymbol{X}_2}{\boldsymbol{X}_1}\right)}{\left(1+\boldsymbol{E}\left(\frac{\boldsymbol{X}_2}{\boldsymbol{X}_1}\right)\right)^3},$$

where $\boldsymbol{E}\left(\frac{\boldsymbol{X}_2}{\boldsymbol{X}_1}\right)$ and $\boldsymbol{Var}\left(\frac{\boldsymbol{X}_2}{\boldsymbol{X}_1}\right)$ are,

$$\boldsymbol{E}\left(\frac{\boldsymbol{X}_2}{\boldsymbol{X}_1}\right)=\frac{\boldsymbol{E}(\boldsymbol{X}_2)}{\boldsymbol{E}(\boldsymbol{X}_1)}\left(1+\boldsymbol{c}^2\right),\ \boldsymbol{Var}\left(\frac{\boldsymbol{X}_2}{\boldsymbol{X}_1}\right)=\left(\frac{\boldsymbol{E}(\boldsymbol{X}_2)}{\boldsymbol{E}(\boldsymbol{X}_1)}\left(1+\boldsymbol{c}^2\right)\right)^2\left(2\boldsymbol{c}^2+\boldsymbol{c}^4\right).$$

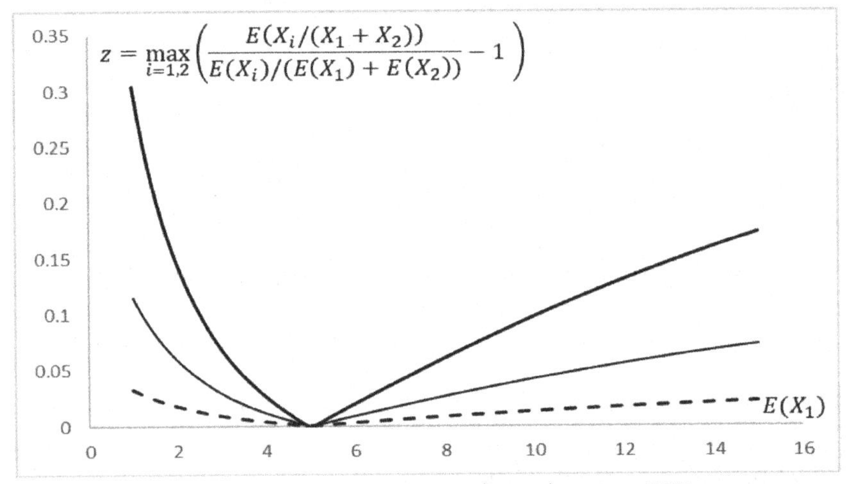

Fig. 2. The maximum relative difference between $E\left(\frac{X_i}{X_1+X_2}\right)$ and $\frac{E(X_i)}{E(X_1)+E(X_2)}$ for X_i, $i=1,2$ distributed *LogNormal*. The bold, thin and dashed lines show the maximum relative difference as a function of $E(X_1)$ for $E(X_2)=5$ and the coefficient of variation of both distributions, $c=1.0, 0.5, 0.25$, respectively.

We observe that $z=0$ at $E(X_1)=E(X_2)=5$ for any c. That is, the lack of fairness disappears if the revenues of the partners are identically distributed, which was to be expected. We observe also that the larger c, the lack of fairness is more pronounced. One can show that the analytical approximation of $E\left(\frac{X_i}{X_1+X_2}\right)$ developed above improves as $E(X_i)$ is increased for $c\le 1$, while $\frac{E(X_i)}{E(X_1)+E(X_2)}$ increases in $E(X_i)$. This explains why z_i is decreasing in $E(X_i)$ as Fig. 2 shows. It is not likely that these properties will change for other values of $E(X_2)$.

In order to overcome the lack of fairness issue, we consider a new allocation scheme and prove that it satisfies the proportionality condition. The allocation scheme is,

$$P_i = \frac{h_i(X_i)}{\sum_{j=1}^{n} h_j(X_j)}, \tag{7}$$

where the increasing functions $h_i(\cdot)$ are defined such that $h_i(X_i)$ will be distributed Gamma for all i with identical scale parameter θ. In other words, the functions $h_i(\cdot)$ transform the given cdf's to cdf's of Gamma. Proposition 1 presents an explicit expression for $h_i(\cdot)$. Note that if the $h_i(\cdot)$'s were allowed to be any increasing functions, then such scheme would not usually satisfy proportionality.

Proposition 1. Let $F_i(\cdot)$ be continuous and strictly monotone on $[0, \infty)\forall i$. Define

$$h_i(\xi) = \left(F_i^{-1}(G_i(k_i, \theta, \cdot))\right)^{-1}(\xi), \tag{8}$$

where superscript (-1) indicates the inverse function, $G_i(k_i, \theta, \cdot)$ is the cdf of the Gamma distribution with scale parameter $\theta > 0$ and shape parameter $k_i = E(X_i)$. Then proportionality holds, i.e.

$$E\left(\frac{h_i(X_i)}{\sum_{j=1}^{n} h_j(X_j)}\right) = \frac{E(h_i(X_i))}{\sum_{j=1}^{n} E\big(h_j(X_j)\big)} = \frac{E(X_i)}{\sum_{j=1}^{n} E\big(X_j\big)} \forall i. \tag{9}$$

Proof. For all strictly increasing $h_i(\cdot)$, the cdf of $h_i(X_i)$ is calculated as follows,

$$Pr(h_i(X_i) \le \xi) = Pr\left(X_i \le h_i^{-1}(\xi)\right) = F_i\left(h_i^{-1}(\xi)\right). \tag{10}$$

By substituting (8) in (10), we obtain,

$$Pr(h_i(X_i) \le \xi) = F_i\left(F_i^{-1}(G_i(k_i, \theta, \xi))\right) = G_i(k_i, \theta, \xi).$$

Thus, the random variable $h_i(X_i)$ is distributed Gamma with shape k_i and scale θ. By substituting $k_i = E(X_i)$, as stated in the proposition, and due to the properties of the Gamma distribution [11 pp. 349–350 and 12, 13] we obtain that

$$E(P_i) = E\left(\frac{h_i(X_i)}{\sum_{j=1}^{n} h_j(X_j)}\right) = \frac{k_i\theta}{\sum_j k_j\theta} = \frac{E(X_i)}{\sum_j E\big(X_j\big)}.$$

■

Example 3. Let the cdf's of the n partners' revenues be identical Weibull distributions, $F(\xi) = 1 - e^{-\frac{\pi\xi^2}{4}}, \xi \ge 0$. The expected revenue of each partner is $E(X) = 1$. The cdf of the Gamma distribution with scale $\theta = 1$ and shape parameter $k = 1$ is $G(1, 1, \xi) = 1 - e^{-\xi}$. The inverse of $F(\xi)$ is

$$F^{-1}(\xi) = \frac{2}{\sqrt{\pi}}\sqrt{-\ln(1-\xi)}.$$

By substituting these expressions in (8), we obtain.

$$h(\xi) = \left(\frac{2}{\sqrt{\pi}}\sqrt{-\ln(1-\left(1-e^{-\xi}\right))}\right)^{-1} = \left(\frac{2}{\sqrt{\pi}}\sqrt{\xi}\right)^{-1} = \frac{\pi\xi^2}{4}.$$

This means that mechanism (7) sets the payment of partner i at $p_i = \frac{\pi x_i^2/4}{\sum_{j=1}^{n}\pi x_j^2/4} = \frac{x_i^2}{\sum_{j=1}^{n}x_j^2}$, where x_j are period's revenues of the partners. According to Proposition 1, $E(P_i) = \frac{E(X_i)}{\sum_j E(X_j)} = \frac{1}{n}$. ■

Scheme (7) satisfies non-negativity and full coverage by definition, monotonicity by the fact that $h_i(\cdot)$ is increasing, and proportionality by Proposition 1.

We also suggest a modification to scheme (5), whose major problem is that partner n's share could be negative in some time periods. The following modification adjusts the partners' shares to be always non-negative. Let the share of partner i, $i = 1, 2, \ldots, n-1$ be

$$p_i = \frac{s_i(x_i)}{\sum_{j=1}^{n}E(X_j)}, \tag{11}$$

and partner $i = n$ completes that to 1, i.e., $p_n = 1 - \sum_{j=1}^{n-1}p_j$. The functions $s_i(\cdot)$, $i = 1, 2, \ldots, n-1$, are mean preserving, i.e., $E(s(X_i)) = E(X_i)$, and the support of $s_i(X_i)$ is the interval $[0, E(X_i) + \frac{1}{n-1}E(X_n)]$. Proposition 2 presents such functions $s_i(\cdot)$.

Proposition 2. Let $F_i(\cdot)$ be continuous and strictly monotone on $[0, \infty)$. Define

$$s_i(\xi) = \left(F_i^{-1}(H_i(\cdot))\right)^{-1}(\xi), \tag{12}$$

where $H_i(\cdot)$ is the cdf of the power distribution.

$$H_i(\xi) = \left(\frac{\xi}{E(X_i) + \frac{1}{n-1}E(X_n)}\right)^{(n-1)E(X_i)/E(X_n)}, \xi \in [0, E(X_i) + \frac{1}{n-1}E(X_n)]. \tag{13}$$

Then, the support of $s_i(X_i)$ is the interval $[0, E(X_i) + \frac{1}{n-1}E(X_n)]$, and $E(s_i(X_i)) = E(X_i)$

.

Proof. For all strictly increasing $s_i(\cdot)$, the cdf of $s_i(X_i)$ is,

$$Pr(s_i(X_i) \le \xi) = Pr\left(X_i \le s_i^{-1}(\xi)\right) = F_i\left(s_i^{-1}(\xi)\right). \tag{14}$$

By substituting (12) in (14), we obtain,

$$Pr(s_i(X_i) \le \xi) = F_i\left(F_i^{-1}(H_i(\xi))\right) = H_i(\xi). \tag{15}$$

That $s_i(X_i)$ is distributed with cdf (13), whose support and mean satisfy the proposition. ■

Example 4. Let $n = 2$ and the cdf's of the partners' revenues be distributed Weibull, $F_i(\xi) = 1 - e^{-\frac{\pi\xi^2}{4a_i^2}}, a_i > 0, \xi \geq 0, i = 1, 2$. The expected revenue of each partner is thus $E(X_i) = a_i$. The cdf of the Weibull distribution is.

$$H_1(\xi) = \left(\frac{\xi}{a_1 + a_2}\right)^{a_1/a_2}, \xi \in [0, a_1 + a_2].$$

The inverse of $F_1(\xi)$ is

$$F_1^{-1}(\xi) = \frac{2a_1}{\sqrt{\pi}}\sqrt{-\ln(1 - \xi)}.$$

.

By substituting these expressions in (15), we obtain

$$s_1(\xi) = \left(\frac{2a_1}{\sqrt{\pi}}\sqrt{-\ln\left(1 - \left(\frac{\xi}{a_1 + a_2}\right)^{a_1/a_2}\right)}\right)^{-1} = (a_1 + a_2)\left(1 - e^{-\frac{\pi\xi^2}{4a_1^2}}\right)^{a_2/a_1}.$$

This means that mechanism (11) sets the payment of partner 1 at $p_1 = \left(1 - e^{-\frac{\pi x_1^2}{4a_1^2}}\right)^{a_2/a_1}$, and partner 2 completes that to 1, $p_2 = 1 - p_1$. By Proposition 2, $0 \leq P_1 \leq 1$ with probability 1, and $E(P_1) = \frac{E(X_1)}{E(X_1)+E(X_2)}$. Accordingly, these hold also for P_2. ■

Scheme (11) satisfies non-negativity and proportionality by Proposition 2, full coverage by definition and monotonicity by the fact that $s_i(\cdot)$ is increasing.

5 Conclusions

Physicians, lawyers and other professional service providers often share an office and other facilities, although they work independently. Such an arrangement, which exists mainly due to economies of scale, raises the question of how to share the office's costs (rent, staff wages) between the parties. Adopting the principle that the allocation has to be by usage as reflected in partners' revenues, which vary over time, we suggested plausible mechanisms of performing such allocation and explore their properties. The mechanisms are based on the partners' shares of total revenue in a given period. What distinguishes among them is whether they use the revenue realizations themselves, functions thereof, or expected revenues.

One might argue that some partners with relatively high revenues might be more efficient than others in their use of facilities, and thus should not be "penalized" by basing allocation on revenues. We disagree for several reasons: a) a major fixed cost is rent which, if use of facilities is not equal, should be reflected in the base allocation (fixed charge); b) in most partnerships the fixed costs will be anyhow significantly lower than revenues; c) monitoring actual usage is time-consuming and difficult[2].

We introduced a mechanism's fairness (proportionality) as the requirement that the long-run average of the partner's share be equal to the ratio of her expected revenue to the sum of the partners' expected revenues. We examine monotonicity and proportionality of the mechanisms and suggest an analytical method which modifies the mechanisms for these properties to hold. The method takes advantage of specific features of the Gamma and Beta distributions.

While the allocation of costs according to a fixed ratio has been widely explored (e.g., [10, 14]), allocating them according to time-varying relative revenues has not been previously investigated. Attempting that raised new issues, which we addressed.

Future research may explore other types of "partnerships" which need to allocate costs. Another direction will be to explore the nature of economies of scale and their implications.

Acknowledgements. The authors thank Ms. Michaella Gerchak for editorial support.

Disclosure of Interests. No funding was received for conducting this study.
The authors declare no competing interests relevant to the contents of this article.

References

1. Young, H.P.: Cost Allocation. In: Handbook of Game Theory with Economic Applications. In: Aumann, R.J., Hart, S. (eds.), vol. 2, pp. 1193–235. Elsevier North Holland, Amsterdam (1994)
2. Carlin, A., Park, R.E.: Marginal Cost Pricing of Airport Runway Capacity. Am. Econ. Rev. **60**(3), 310–319 (1970)
3. Littlechild, S.C., Thompson, G.F.: Aircraft landing fees: a game theory approach. Bell J. Econ. **8**(1),186–204 (1977). https://doi.org/ https://doi.org/10.2307/3003493
4. O'Neill, B.: A problem of rights arbitration from the Talmud. Math. Soc. Sci. **2**(4), 345–371 (1982). https://doi.org/ https://doi.org/10.1016/0165-4896(82)90029-4
5. Aumann, R.J., Maschler, M.: Game theoretic analysis of a bankruptcy problem from the Talmud. J. Econ. Theory **36**(2),195–213 (1985). https://doi.org/ https://doi.org/10.1016/0022-0531(85)90102-4
6. Thomson, W.: Axiomatic and game-theoretic analysis of bankruptcy and taxation problems: an update. Math. Soc. Sci. **74**, 41–59 (2015). https://doi.org/ https://doi.org/10.1016/j.mathsocsci.2014.09.002
7. Ertemel, S., Kumar, R.: Proportional rules for state contingent claims. Inter. J. Game Theory **47**(1),229–46 (2018). https://doi.org/ https://doi.org/10.1007/s00182-017-0585-7
8. Xue, J.: Fair division with uncertain needs. Soc. Choice Welfare **51**(1), 105–36 (2018). https://doi.org/ https://doi.org/10.1007/s00355-018-1109-5
9. Guajardo, M., Rönnqvist, M.: A review on cost allocation methods in collaborative transportation. Inter. Trans. Operational Res. **23**, 371–92 (2016). https://doi.org/ https://doi.org/10.1111/itor.12205

[2] Not surprisingly, many partnerships just allocate fixed costs at a constant ratio, even equally, regardless of actual usage of facilities in that period.

10. Billera, L.J., Heath, D.C.: Allocation of shared costs: a set of axioms yielding a unique procedure. Math. Oper. Res. **7**(1), 32–39 (1982)
11. Johnson, N.L., Kotz, S., Balakrishnan, N.: Continuous Univariate Distributions, vol. 1, 2nd ed. Wiley, New York (1994)
12. Wikipedia: Dirichlet distribution. https://en.wikipedia.org/wiki/Dirichlet_distribution, (Accessed 19 Jan 2022)
13. Wikipedia: Gamma distribution. https://en.wikipedia.org/wiki/Gamma_distribution, (Accessed 19 Jan 2022)
14. Young, H.P.: Cost Allocation: Methods, Principles. Applications. Elsevier North Holland, Amsterdam (1985)

Examining the Effect of ChatGPT on Small Group Ideation Discussions

Madoka Chosokabe(✉), Shohei Koie, and Yuji Oyamada

Tottori University, 4-101 Koyama Minami, Tottori, Japan
mchoso@tottori-u.ac.jp

Abstract. In discussions among residents, such as those in town planning workshops, participants often engage in conversations and propose ideas for community development based on given themes. However, due to differences in knowledge about the themes among participants, establishing a common understanding takes time, and there is a limited timeframe for considering and consolidating ideas. As a result, facilitators play a crucial role in guiding the discussions among residents, but there is a shortage of facilitators relative to the opportunities for discussions. Consequently, it is challenging for participants alone to consolidate high-quality ideas. Generative AI tools like ChatGPT are anticipated to be valuable in supporting idea generation. On the other hand, there is a risk of developing ideas overly dependent on generative AI if used improperly. Therefore, it is crucial to first understand the effect that generative AI can have on group discussions. In this paper, we conduct a user study to examine effective usages of ChatGPT during small group ideation discussions. From this experiment, we discovered and categorized both the positive and negative effects from utilizing ChatGPT.

Keywords: Participatory Planning Process · Small Group Discussion · ChatGPT

1 Introduction

Several small group discussions among community residents are held during a participatory planning process in order to make better decisions, incorporate their opinions into the local governments' planning, and to produce ideas for the betterment of the entire community. In the future, there will be an increasing need within communities to engage in creative discussions—not just for consensus-building discussions. For example, citizen-led creative discussions would be crucial in generating ideas for new mechanisms and projects aimed at achieving sustainable local communities [1]. Here, an idea is defined as a new combination of existing knowledge. Specifically, an idea is generated from the following five processes: 1) the collection of materials; 2) digestion; 3) combination; 4) the birth of an idea; and 5) materialization [2]. Group idea exchange in organizations can improve with attention and reflection. "Brainwriting" study suggests enhanced sharing leads to better idea generation, fostering creativity and innovation [3].

M. Campos Ferreira et al. (Eds.): GDN 2024, LNBIP 509, pp. 142–153, 2024.
https://doi.org/10.1007/978-3-031-59373-4_12

In a small group discussion, diverse knowledge possessed by various participants is expected to both diverge and converge into a single idea by successfully combining them. However, there are two issues with small group discussions for generating ideas. First, there are differences in residents' knowledge of any given subject matter—and second, there is a lack of sufficient time for participants to adequately deepen their mutual understanding and refine ide. Regarding knowledge gaps, the findings of Palermo and Hernandez [4] highlighted barriers in participatory adaptation planning, including communication and knowledge gaps, as well as challenges in integrating and coordinating stakeholders. For the former, several attempts have been made—including lectures by experts that help to further explain the subject matter. For the latter, it is difficult to reach the point of materializing ideas in a limited amount of time for the discussion.

To enhance creative discussion, the facilitator—who assists participants in smoothly discussing with each other—is required to have high level skills in order to organize and consolidate opinions while mitigating the differences in participants' knowledge [5]. In our previous research [6], we focused on small group discussions in Japanese local government planning. We highlighted the importance of facilitation skills, especially during the strategy proposal phase, for effective idea generation. Our findings show that the number of ideas varies based on the discussion topic and phase, with some engaging all participants and others only a few. This diversity also applies to participants' knowledge levels on the topic. However, despite the importance of facilitators in small group discussions, due to a shortage of human resources, it is unlikely that there are enough highly skilled facilitators within the community.

ChatGPT, based on GPT-3.5, has significantly impacted non-engineers with its human-like chat capabilities, encouraging widespread interaction and experimentation. Its authenticity in replies and reactions reduces user hesitation, fostering extensive sharing of experiences on social media, thus accelerating its growth. ChatGPT has been explored for its application in decision-making and ideation in various fields—such as education and product development—since its public release in November 2022. The potential and limitations of AI creativity are still under discussion. Haase and Hanel [7] compared human-generated ideas with those from six AI chatbots, finding no qualitative difference in creativity. It suggests that AI can assist in the creative process, but further research is needed to understand its full potential and limitations. They also question whether AI can achieve "true" creativity. Ivcevic and Grandinetti [8] focuses on co-creation, where AI augments human creativity. They discuss research on AI's role in creativity across four levels: mini-c/creativity in learning, little-c/everyday creativity, Pro-C/professional creativity, and Big-C/eminent creativity.

Research on the utilization of ChatGPT in the field of education includes the following examples. Chauncey and McKenna [9] explored how AI chatbots, like ChatGPT, could ethically enhance critical thinking and learning in education. Mogavi et al. [10] examined ChatGPT's impact on education through social media analysis. It was widely used in higher education and K-12 settings, with discussions focusing on productivity and ethical concerns. Users' views varied, highlighting the complexity of integrating AI in education. Basic et al. [11] compared essay writing with and without ChatGPT-3 assistance. Results showed no significant improvement in essay quality with ChatGPT-3.

AI improves academic writing by offering tools for grammar checks, plagiarism detection, and language translation, though concerns remain about its impact on creativity and ethical writing practices [12]. Exintaris et al. [13] used ChatGPT in a problem-solving workshop to stimulate metacognition among students regarding chemistry problems. Students identified errors in ChatGPT's solutions, demonstrating the effectiveness of metacognition in problem-solving. Essel et al. [14] examined the impact of using ChatGPT on university students' critical, creative, and reflective thinking skills in Ghana. It utilized a mixed-methods approach with 125 students, showing significant improvement in these skills when ChatGPT was integrated into the class. Yilmaz and Yilmaz [15] evaluated ChatGPT's effect on programming education. The experimental group, using ChatGPT during programming tasks, showed improved skills and motivation compared to the control. This underscores ChatGPT's potential in programming training.

ChatGPT can potentially address earlier mentioned issues, residents have varying knowledge levels, and time constraints limit idea development during discussions, by bridging knowledge gaps among participants. Instead of individuals researching and restructuring understanding, ChatGPT handles these steps, enhancing efficiency. However, occasional inaccuracies require consideration for reply accuracy. Additionally, there may be negative effects on participants' creative discussions. Habib et al. [16] examined how generative AI affects students' creative thinking. Findings caution against quick AI integration in creative education due to potential negative effects on creativity and confidence.

Using ChatGPT responses to initiate discussions allows for broader exploration of topics, potentially easing time constraints. However, in small ideation discussions, the potential of ChatGPT for supporting our creation is unclear. The aim of this study is to examine effective usages of ChatGPT within small group ideation discussions. For this purpose, we conduct a subject experiment that compares group discussion with and without ChatGPT. We analyze the differences between discussions and examine both the pros and cons of ChatGPT utilization.

2 Experimental Outline

The overview of the experiment is presented in Table 1. 36 engineering students in Tottori University participated, with each discussion session including three subjects and one ChatGPT operator using version 4.0. To account for potential variations, two topics were chosen: one familiar to Tottori University students, focusing on congestion alleviation in shops and cafeterias, and another less familiar, aiming to increase tourism in Tottori Prefecture. Additionally, different question card types and usage patterns were implemented to clarify ChatGPT's role, resulting in four conditions. The experimental procedure included three subjects discussing Topics 1 and 2 under the same conditions on separate days. Each group underwent three trials for each of the four conditions. Data collection methods included images, audio recordings, videos, transcripts, the 6W2H framework, and surveys to document the discussion process. The experiment employs question cards to assess ChatGPT's effect on discussions and ideas, categorized into fact-based and open-ended types. Fact-based cards yield consistent answers and cover facts, definitions, methods, causes, and effects. Open-ended cards prompt diverse responses, focusing on advice, opinions, and recommendations (See also Table 2). In the experiment, subjects summarize discussion ideas using a 6W2H framework, capturing key aspects comprehensively. This framework covers who, to whom, what, how, when, where, how much, and why for each idea, ensuring clarity. Subjects are instructed to be specific, avoid lengthy responses, use bullet points, and provide unbiased, comprehensive information. These guidelines are provided beforehand to facilitate completion of the framework. Three subjects participate in a 40-min discussion on the topic provided. Each group presents one idea and condenses it into a single 6W2H framework. Subjects in Conditions 1–3 have the option to use predefined question cards. ChatGPT is operated by the operator, and subjects communicate the questions they wish to ask using expressions specified in the question cards.

Table 1. Experiment outline

Participants	For each group, there are three subjects and one ChatGPT operator
Topics	1. Alleviating congestion in cooperative shops and cafeterias; familiar topic to Tottori University students 2. Generating ideas to increase tourists in Tottori Prefecture; less familiar topic to Tottori University students
Methods	The same three subjects (one group) discuss Topic 1 and Topic 2 under identical conditions. Subsequently, they consolidate ideas onto a single 6W2H framework
Conditions	1. Subjects can use both fact-based and open-ended question cards 2. Subjects can use only fact-based question cards 3. Subjects can use only open-ended question cards 4. Subjects cannot use ChatGPT

Table 2. Two types of question cards

Type	Definition	Example
Fact-based	Facts/Definitions	Please tell me about X
	Methods/Means	Can you teach me how to X?
	Causes/Reasons	Why is it X?
	Effects/Results	What happens when I X?
Open-ended	Advice	What should I do about X?
	Opinion	What do you think about X?
	Recommendation	Can you recommend a good X?

3 The Survey Results for the Subjects

3.1 Evaluations of Discussion and Idea

Subjects evaluated their discussion on the following seven aspects—using a 5-point scale that ranges from agree, agree a little, neutral, disagree a little, to disagree:

1. I was able to express my thoughts adequately.
2. We could generate a lot of ideas.
3. I gained new knowledge.
4. I could share opinions with others.
5. The atmosphere of the discussion was positive.
6. The discussion progressed smoothly.
7. The discussion reached a consensus.

Similarly, they evaluated the ideas generated by the group on the following five aspects—using a 5-point scale ranging from agree, agree a little, neutral, disagree a little, to disagree:

1. The ideas became innovative.
2. The ideas became specific.
3. The ideas became creative.
4. The ideas became feasible.
5. The ideas became useful.

The aggregated results for the evaluation of discussions in the questionnaire items related to discussions are presented in Fig. 1. Focusing on the response "Agree," it is observed that, when excluding the opinion sharing item, the percentage is larger for discussions without ChatGPT than with ChatGPT for all other questionnaire items; in particular, the proportion is significant for smoothness. Therefore, there is a possibility that the use of ChatGPT may impose a burden on subjects' discussions. However, no statistically significant differences were observed in any of the questions between the two groups.

The aggregated results for the evaluation of ideas related to ideas are presented in Fig. 2. Focusing on the response "Agree," it is observed that, for all questionnaire items, the percentage is larger for discussions without ChatGPT than with ChatGPT. However, when considering the combined percentages of "Agree" and "Agree a little," it was found that the percentage is larger when ChatGPT is used only for feasibility.

Additionally, the difference in the combined percentages between ChatGPT-present and ChatGPT-absent is most notable for originality. Therefore, from the subjects' perspective, it is possible that the use of ChatGPT may lead to more realistic ideas but may result in fewer creative ideas. However, no statistically significant differences were observed in any of the questions between the two groups.

3.2 Evaluations of ChatGPT

The subjects using ChatGPT for conditions 1–3 evaluated the utility of ChatGPT during discussions on the following 11 aspects—using a 5-point scale ranging from agree, agree a little, neutral, disagree a little, to disagree:

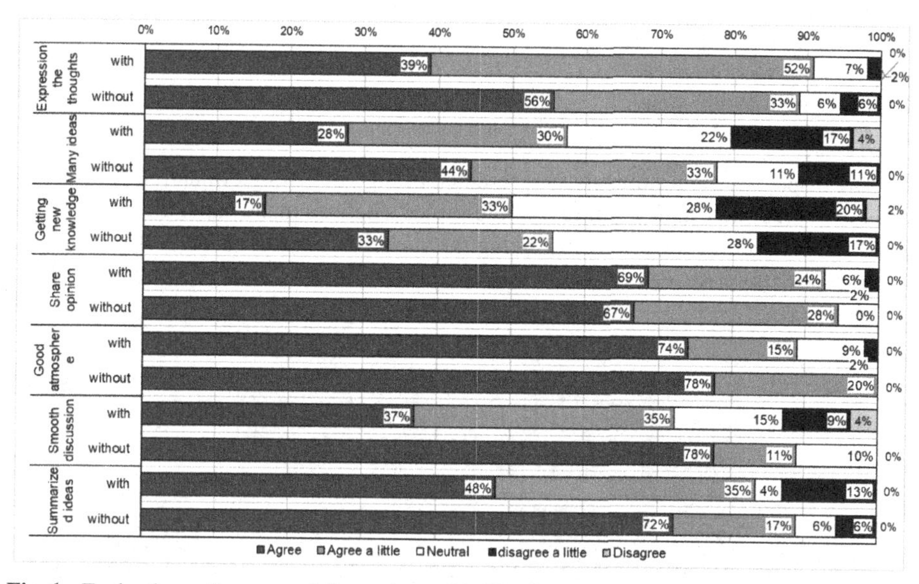

Fig. 1. Evaluations of process of discussion: with ChatGPT (n = 54) and without ChatGPT (n = 18), with a total of 36 subjects providing two responses at each discussion.

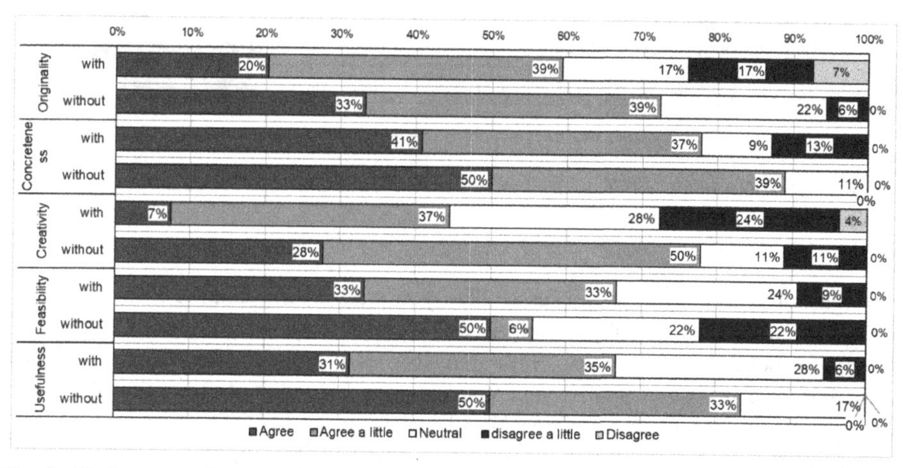

Fig. 2. Evaluations of summarized ideas: with ChatGPT (n = 54) and without ChatGPT (n = 18), with a total of 36 subjects providing two responses at each discussion.

1. Gaining new knowledge.
2. Sharing opinions with other subjects.
3. Improving the atmosphere of the discussion.
4. Smoothly progressing the conversation.
5. Summarizing everyone's opinions.
6. Generating numerous ideas.
7. Creating new ideas.

8. Creating innovative ideas.
9. Creating specific ideas.
10. Creating feasible ideas.
11. Creating useful ideas.

The responses of the subjects are presented in Fig. 3—listing evaluation items in the order of "agree" and "agree a little." While subjects believe that ChatGPT contributes to generating numerous ideas and enhancing the specificity and usefulness of ideas, there are negative responses regarding its ability to generate creative ideas or summarize ideas.

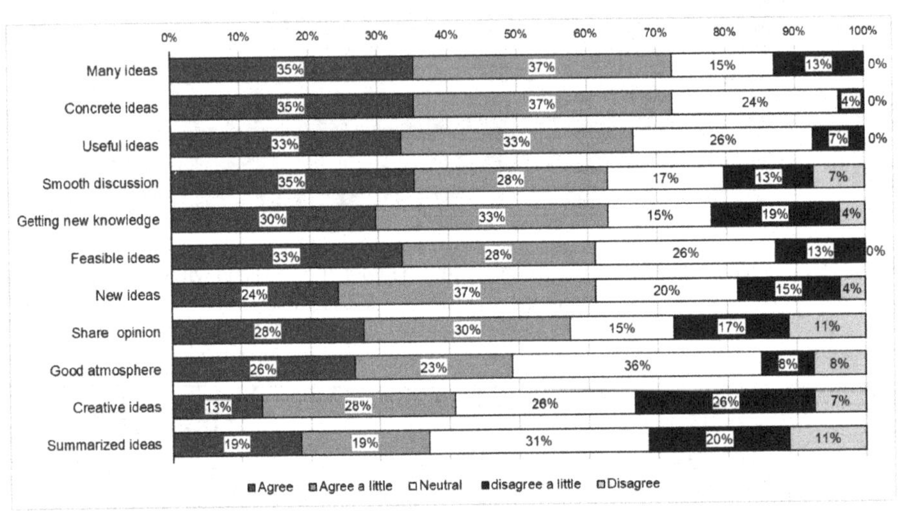

Fig. 3. Evaluations of ChatGPT: with ChatGPT (n = 54) and a total of 27 subjects providing two responses at each discussion.

Figure 4 presents the responses to "Do you want to use ChatGPT in future discussions?" with 78% of subjects expressing a positive answer. Figure 5 categorizes the reasons behind each response, with analysts classifying them into eight categories: Smooth Progress, Ease of Sharing and Visualization, Immediate Information Provision, Enrichment and Presentation of Ideas, Novelty of Topics and Idea Stimulation, Concerns About Reliability, Adaptability to ChatGPT, and Limited Usage. The subjects' specific feedback is discussed in Sect. 3.3.

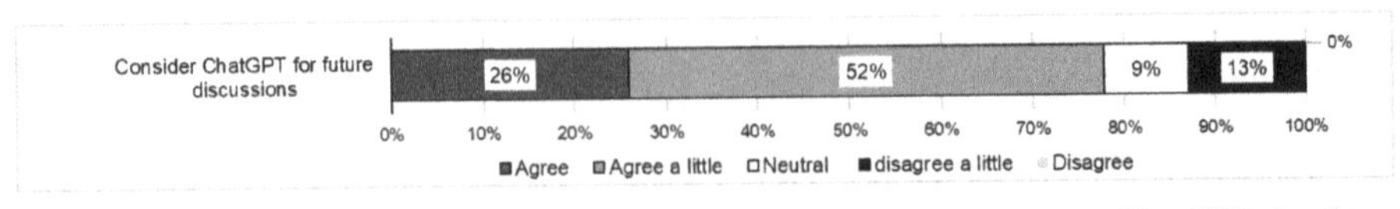

Fig. 4. The response of the subject to the question: Do you want to use ChatGPT for future discussions?

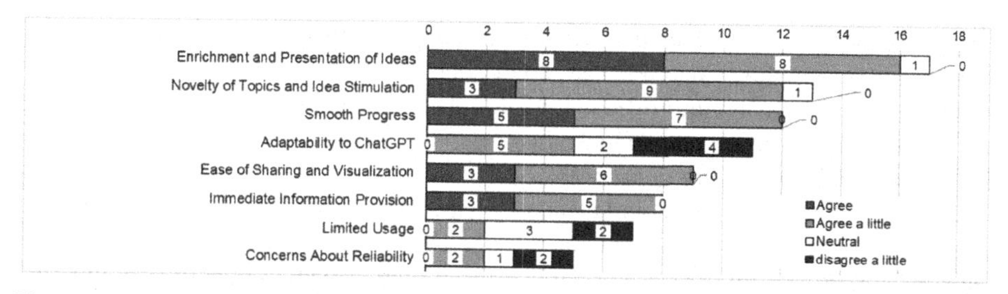

Fig. 5. Results of categorizing the opinions of subjects in response to the question, "Do you want to use ChatGPT for future discussions?" with a total of 27 subjects providing two responses at each discussion.

3.3 Subjects' Feedback on Using ChatGPT in Discussions

Positive Feedback.
Figure 6 summarizes subjects' feedback on the question "Tell us what was good about discussions using ChatGPT." Subjects provided their opinions after the first and second experiments. We categorized their feedback into four themes: support in facilitating discussions, idea generation and proposal, provision of new information and knowledge expansion, and support in finding solutions. "Topic1" refers to the results of subjects' responses after the first experiment. An overview of each opinion is presented below.

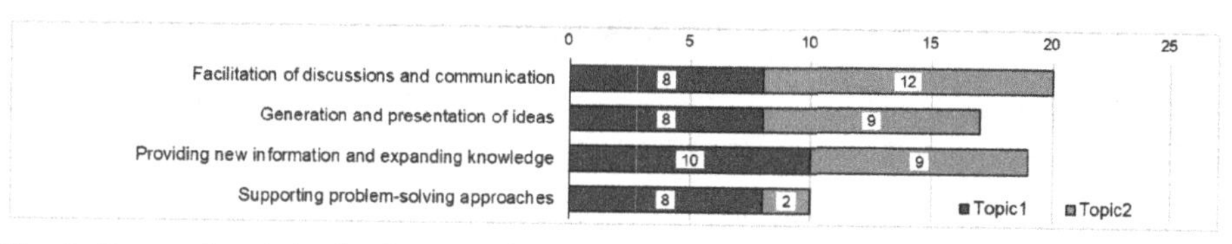

Fig. 6. Respondents' feedback on positive aspects of ChatGPT (n = 52)—with a total of 26 subjects providing two responses at each discussion.

Facilitation of discussions and communication.
ChatGPT facilitated discussions by providing a shared focus, smooth progression, and expert support. It enhanced communication by making it easier to express opinions, maintain a conducive atmosphere, and reduce unnecessary time. Additionally, it contributed to expanding the conversation, determining its direction, and promoting diverse opinions while maintaining appropriate limitations.
Generation and presentation of ideas.
ChatGPT greatly enhanced discussions by providing diverse and abundant ideas, facilitating decision-making, introducing unique perspectives, and swiftly generating creative suggestions. It guided conversations with humorous title suggestions, overcame idea blocks, and positively received new opinions—contributing to innovative thinking and enhanced persuasiveness.
Providing new information and expanding knowledge.
In discussions using ChatGPT, there were diverse answers, effective suggestions for solutions, and a valuable provision of new information and ideas. The reassuring and precise

responses, along with new perspectives, enhanced the depth of discussions and assisted in organizing thoughts. ChatGPT supported congestion relief, introduced new topics, enriched vocabulary, and improved knowledge about regions and themes—contributing to smooth and productive discussions.

Supporting problem-solving approaches.

In ChatGPT discussions, we could evaluate ideas, receive diverse suggestions for solutions and titles, and benefit from the convenience of quick responses—helping to reduce the need for online searches. The provided hints and ideas effectively supported problem-solving and enriched our discussions.

Negative Feedback.

Figure 7 summarizes the subjects' negative feedback. We categorized their feedback into five themes: poor quality of responses, lack of specificity, low accuracy and reliability of information, time-consuming, inhibition of communication due to difficulty in usage, and doubts about compatibility with ChatGPT in discussions. An overview of each opinion is presented below.

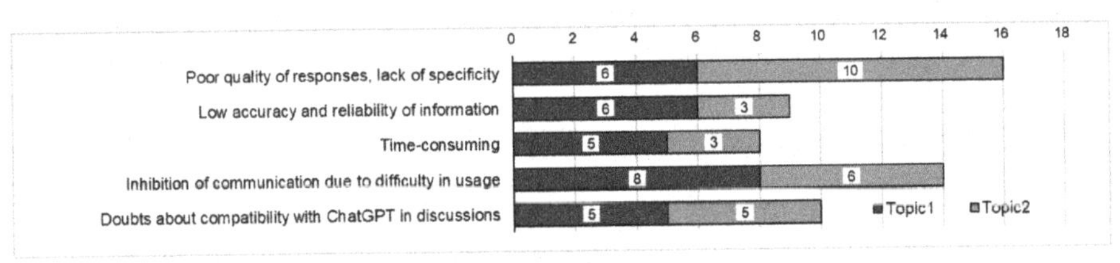

Fig. 7. Respondents' feedback on negative aspects of ChatGPT (n = 54)—with a total of 27 subjects providing two responses at each discussion.

Poor quality of responses, lack of specificity.

The drawbacks of using ChatGPT in discussions were related to the quality of responses. There were varying answers to specific numbers or methods; furthermore, explanations for solutions were lengthy and lacked conciseness. For certain topics, the use of ChatGPT felt unnecessary, and responses were often general and lacking innovation. Specific information or original ideas were not always obtained, posing challenges when expected responses were not received.

Low accuracy and reliability of information.

There is a possibility of misinformation or falsehoods, as information from ChatGPT may not always be accurate—leading to discussions based on incorrect information. The lack of consistency in answers to questions and the difficulty in verifying accuracy were sources of dissatisfaction. The information provided is limited to that up to 2021; moreover, concerns were raised about the inability to expect new knowledge and the difficulty in detecting false answers.

Time-consuming.

The conversation changed with the responses, which lead to waiting times. The uncertainty of obtaining answers for each question required time for consideration and articulation, and especially with lengthy responses—the waiting time felt inefficient. As a result, inconveniences such as interruptions in discussions, a slowed pace in conversation, and the need to revisit topics occurred.

Inhibition of communication due to difficulty in usage.

Using ChatGPT hindered communication due to challenges in precise wording and potential for unexpected responses, especially in open-ended type questions. The difficulty in incorporating suggestions into ideas, reliance on ChatGPT for decision-making, and the challenge of questioning without a clear group direction were also observed. There was a concern about being influenced by ChatGPT's suggestions.

Doubts about compatibility with ChatGPT in discussions.

ChatGPT usage raised concerns about its compatibility with discussions. Subjects found the generated ideas to be largely conventional and not creatively engaging. The choice of themes and questioning methods, coupled with the perception of limited originality in responses, made it challenging to leverage ChatGPT effectively. Concerns were also raised about the risk of overreliance on the program, leading to a lack of innovation and the need for a balanced approach in utilizing ChatGPT in discussions.

Lessons Learned from the Experiment.

Many skeptics of using ChatGPT in discussions worry about its usability and information reliability. While unfamiliarity may pose an initial challenge, participants can adapt with practice or skilled operators. In diverse community discussions, some may lack familiarity with ChatGPT, but facilitators could manage its use. To address concerns about information reliability, integrating ChatGPT with more credible tools, like internet searches, could help.

Subjects who have a positive view of utilizing ChatGPT in discussions acknowledge that ChatGPT swiftly offers diverse ideas, aids in decision-making, and encourages new perspectives. Additionally, it provides reassurance, helps overcome idea blocks, and enhances discussions by offering valuable insights and solutions. However, some participants expressed concerns about becoming dependent on ChatGPT's ideas. To prevent overreliance on ChatGPT in discussions, it may be advisable for facilitators to operate it. Given that community meetings often involve individuals who are unfamiliar with each other, creating a comfortable atmosphere as quickly as possible is crucial. Similarly, participants also valued the ability to share common perspectives thanks to ChatGPT. This would be the case if facilitators adeptly operate ChatGPT, surpassing its delayed responses and uncertain information, resulting in positive outcomes.

4 Summary

Positive views on ChatGPT usage included easier visualization of opinions, smoother discussions, and its utility for refreshing during intense moments. Generating multiple ideas, lively discussions, filling knowledge gaps, and broadening the discussion scope were noted as positive views. On the other hand, although neutral or negative views mentioned the usefulness for creative aspects, they cautioned about potential inaccuracies; furthermore, while it facilitated smooth discussions, some doubted its effectiveness in acquiring desired information. Subjects suggested specific use cases and expressed concerns about topic dependence and limitations in providing valuable insights. Some found it cumbersome due to unfamiliarity, while others recommended traditional search methods for fact-checking. Overall, subjects acknowledged both the benefits and challenges in integrating ChatGPT into discussions.

During the participatory planning process, at the small group discussion, it is difficult for subjects to generate specific ideas because of both the participants' knowledge and time limitation. This paper aimed to examine effective usages of ChatGPT within small group ideation discussions. We conducted a subject experiment to compare group discussions with and without ChatGPT. From the subjects' feedback and the observers' impressions, we confirmed the positive and negative effects of ChatGPT usage. In particular, it may be advisable to decide in advance among both facilitators and participants how ChatGPT will be utilized. To maximize the positive effects while minimizing the negative ones, we need to conduct further experiments with a broader—and more diverse—pool of participants.

Acknowledgments. This study was funded by JSPS KAKENHI Grant Number JP21K14267, 24K07713, the Platform for Community-based Research and Education (CoRE) and Center for Regional Management and Safety Engineering (CRMSE), Tottori University.

Disclosure of Interests. The authors have no competing interests to declare that are relevant to the content of this article.

References

1. Leminen, S., Rajahonka, M., Westerlund, M., Hossain, M.: Collaborative innovation for sustainability in Nordic cities. J. Clean. Prod. **328**, 129549 (2021). https://doi.org/10.1016/j.jclepro.2021.129549
2. Young, J.W.: A Technique for Producing Ideas (Revised Edition). McGraw-Hill (2003)
3. Paulus, P.B., Yang, H.-C.: Idea generation in groups: a basis for creativity in organizations. Organ. Behav. Hum. Decis. Process.Behav. Hum. Decis. Process. **82**, 76–87 (2000). https://doi.org/10.1006/obhd.2000.2888
4. Palermo, V., Hernandez, Y.: Group discussions on how to implement a participatory process in climate adaptation planning: a case study in Malaysia. Ecol. Econ. **177**, 106791 (2020). https://doi.org/10.1016/j.ecolecon.2020.106791
5. Rees, F.: The Facilitator Excellence Handbook. Wiley, San Francisco (2005)
6. Chosokabe, M., Tanimoto, K., Tsuchiya, S.: Design of small group discussion using the framework of business process management: the case of Japanese participatory planning process. In: Proceedings of the 2020 Group Decision and Negotiation (2020)
7. Haase, J., Hanel, P.H.P.: Artificial muses: generative artificial intelligence chatbots have risen to human-level creativity. J. Creativity. **33**, 100066 (2023). https://doi.org/10.1016/j.yjoc.2023.100066
8. Ivcevic, Z., Grandinetti, M.: Artificial intelligence as a tool for creativity. J. Creativity. **34**, 100079 (2024). https://doi.org/10.1016/j.yjoc.2024.100079
9. Chauncey, S.A., McKenna, H.P.: A framework and exemplars for ethical and responsible use of AI chatbot technology to support teaching and learning. Comput. Educ. Artif. Intell. **5**, 100182 (2023). https://doi.org/10.1016/j.caeai.2023.100182

10. Mogavi, R.H., et al.: ChatGPT in education: a blessing or a curse? A qualitative study exploring early adopters' utilization and perceptions. Comput. Hum. Behav. Artif. Hum. **2**, 100027 (2024). https://doi.org/10.1016/j.chbah.2023.100027
11. Basic, Z., Banovac, A., Kruzic, I., Jerkovic, I.: ChatGPT-3.5 as writing assistance in students' essays. Humanit. Soc. Sci. Commun. **10**, 750 (2023). https://doi.org/10.1057/s41599-023-02269-7
12. Malik, A.R., et al.: Exploring artificial intelligence in academic essay: higher education student's perspective. Int. J. Educ. Res. Open. **5**, 100296 (2023). https://doi.org/10.1016/j.ijedro.2023.100296
13. Exintaris, B., Karunaratne, N., Yuriev, E.: Metacognition and critical thinking: using chatGPT-generated responses as prompts for critique in a problem-solving workshop (SMARTCHEMPer). J. Chem. Educ. **100**, 2972–2980 (2023). https://doi.org/10.1021/acs.jchemed.3c00481
14. Essel, H.B., Vlachopoulos, D., Essuman, A.B., Amankwa, J.O.: ChatGPT effects on cognitive skills of undergraduate students: receiving instant responses from AI-based conversational large language models (LLMs). Comput. Educ. Artif. Intell. **6**, 100198 (2024). https://doi.org/10.1016/j.caeai.2023.100198
15. Yilmaz, R., Yilmaz, F.G.K.: The effect of generative artificial intelligence (AI)-based tool use on students' computational thinking skills, programming self-efficacy and motivation. Comput. Educ. Artif. Intell. **4**, 100147 (2023). https://doi.org/10.1016/j.caeai.2023.100147
16. Habib, S., Vogel, T., Anli, X., Thorne, E.: How does generative artificial intelligence impact student creativity? J. Creativity. **34**, 100072 (2024). https://doi.org/10.1016/j.yjoc.2023.100072

Full Rank Voting: The Closest to Voting with Intensity of Preferences

Luis G. Vargas[1(✉)] and Marcel C. Minutolo[2]

[1] University of Pittsburgh, Pittsburgh, PA, USA
lgvargas@pitt.edu
[2] Robert Moris University, Moon, PA, USA

Abstract. Voting is a fundamental aspect of democracy, but traditional voting schemes often fail to capture the intensity of preferences individuals possess. This loss of intensity can lead to an oversimplification of complex issues and a lack of accurate representation of diverse opinions. To address this limitation, we propose a voting method called "Full Rank Voting" that incorporates the intensity of preferences into the voting process. By using pairwise comparisons and Saaty's fundamental scale, we transform individual preferences into numerical values and construct a matrix that represents the intensity of preferences for different candidates or options. In the case of two candidates, each voter expresses their preference intensity by assigning a numerical value from the fundamental scale. These values are then used to calculate the priorities or percentages of votes for each candidate. By incorporating intensity of preferences, the voting process becomes more nuanced, and ordinal preferences become a specific case of cardinal preferences. When multiple candidates are involved, we encounter the challenge of combining intensity of preferences with rank voting. We conduct simulation experiments to demonstrate that rank voting and voting with intensity of preferences yield similar results, even for relatively small sample sizes. Overall, Full Rank Voting offers a solution for capturing the intensity of preferences in voting, leading to a more accurate representation of individual choices, increased democratic legitimacy, and the ability to identify common ground and prioritize preferences based on their strength.

Keywords: Voting · Rank · Comparisons · Preferences · Intensity

1 Introduction

Voting is the basis of democracy; one person one vote. However, not everybody likes (dislikes) a candidate with the same intensity though this is lost in most voting schemes. It is important for a democracy to capture the intensity of preferences since knowing it provides a more accurate representation of choice and more nuanced understanding of individual preferences. There are many reasons why it is important for a democracy to capture intensity of preferences when voting. For instance, it reflects diversity of opinion. By capturing intensity, we can recognize the varying degrees of support or opposition individuals may have toward different options. This ensures that the diversity of opinions within a population is accurately represented.

M. Campos Ferreira et al. (Eds.): GDN 2024, LNBIP 509, pp. 154–170, 2024.
https://doi.org/10.1007/978-3-031-59373-4_13

Voting is not the same as group decision making that may require achieving a consensus. To achieve a consensus, if the intensity of preferences is represented using a numerical scale, the numerical preferences may be required to be close to each other so that the synthesis represents group's preference. Voting does not require synthesis of numerical preferences, just account for the vote in whatever form it is provided.

When we capture the intensity of preferences, we enhance democratic legitimacy. Since a democracy seeks to reflect the will of the people, when intensity of preferences is considered, we get a more accurate measurement of collective will. Knowing more accurately the collective will helps to increase the legitimacy of the outcomes, as they align more closely with the true sentiments of the electorate. Treating all preferences equally without considering their intensity can lead to oversimplification of complex issues; it fails to account for the strength of convictions individuals have towards specific choices. By capturing intensity, we gain a deeper understanding of the underlying motivations and beliefs behind people's preferences.

Additionally, a more nuanced understanding facilitates compromise and consensus-building. When the intensity of preferences is captured, it becomes easier to identify common ground and areas of compromise (e.g., Saaty et al., 2022). Policymakers and decision-makers better understand the trade-offs and preferences of different groups when they take into consideration the intensity of preferences, which can lead to more inclusive and balanced outcomes. Further, decision-makers are able to better prioritize preferences based on how strongly or not various outcomes are weighted. Some issues may be more important to individuals than others. By capturing the intensity of preferences, it becomes possible to prioritize issues or policies based on the level of support or opposition they generate. This helps in allocating resources and attention to the most critical concerns of the population. The motivation behind this work is to offer a solution where intensity of preferences is retained in voting for improved democracy. So, how do we incorporate into the voting process the intensity with which we feel our support/like/dislike for a candidate? To begin, we must first explain what intensity of preference means.

To represent intensity of preferences we need to be able to measure our preferences in a scale. We usually use words to represent how strongly we prefer something, e.g., I like it a lot, very much, not all, and so on. These words need to be associated with numbers so that we can combine the preferences to represent how something is liked or preferred. We could count how many people in a group prefer something strongly, but counting does not represent how strongly two individuals prefer the item or the candidate. To be able to represent intensity numerically we need to use measurement. One could say a person prefers A to B strongly, and another could say that she prefers A to B equally. We cannot mathematically combine the intensities "strongly" and "equally" because they belong to a nominal scale. However, if they were to be assigned a numerical value that satisfy some ordinal condition, e.g., the number assigned to "strongly" must be greater than the number assigned to "equally," we could try to transform the entire set of preferences of a group into a numerical value. Clearly, those numerical assignments may not be the same for everybody in a group. Because that scale may not be the same for everyone, we may never agree on the definition of the unit of measurement. Thus, we need relative measurement which does not use units; a simple example can explain this.

There is another school of thought which models preferences by linguistic preference relations with fuzzy sets e.g. (Zadeh, 1975a); (Zadeh, 1975b); (Zadeh, 1975c); (Herrera et al., 2000); (Herrera-Viedma et al., 2005); and, (Zhou et al., 2008).

Consider a set of stones that we need to rank in terms of weight (Saaty & Vargas, 2007), but we do not have a gadget to ascertain their weight. We can take the stones in our hands and guess which one is heavier, but to do that we need to select them in pairs and order them in terms of the perceived weight. However, ordering them does not allow us to find their relative weights. To find the relative weights, we need to assign to each pair comparison a numerical value that somehow reflects our perception of weights, e.g., one stone is heavier than the other and, in this case, we think that one is 3 times heavier than the other. This would mean that the total weight of three stones the size of the smaller stone would be equal to the weight of the larger stone. Thus, we need to build a scale that maps the words representing intensity to numbers. This subject falls under the disciple known as psychophysics (Fechner, 1966). Gustav Theodor Fechner, a German philosopher and scientist, developed the area of psychophysics. Psychophysics investigates the quantitative associations between mental and physical phenomena, focusing on the precise correlation between sensory perceptions and the external stimuli generating them. (Saaty, 1980) developed his fundamental 1–9 scale using psychophysics principles. In what follows, we express numerical preferences by Saaty's 1–9 scale.

2 What is Voting with Intensity of Preference

2.1 Two Candidates

When voting, a person makes a choice between, for example, two candidates. The choice is a 0,1 decision. Assume that we could vote by also providing how strongly we feel about the candidate. Thus, we could say I prefer candidate A to candidate B strongly. We need to assign a numerical value to the intensity "strong." We could use Saaty's fundamental scale given in Table 1.

With two candidates, A and B, we can arrange the judgment expressing intensity of preference in a matrix:

$$\begin{array}{c} \\ A \\ B \end{array} \begin{array}{c} A \; B \\ \begin{pmatrix} 1 & a \\ 1/a & 1 \end{pmatrix} \end{array}$$

The value a is a number from the fundamental 1–9 scale assigning the intensity of preference for one candidate over another. The priorities of the candidates for that judgment are the normalized to unity values of any of the columns of the matrix:

$$\begin{array}{c} \\ A \\ B \end{array} \begin{array}{c} A \; B \\ \begin{pmatrix} 1 & a \\ 1/a & 1 \end{pmatrix} \end{array} \begin{array}{c} \text{Priorities} \\ \begin{pmatrix} a/(a+1) \\ 1/(a+1) \end{pmatrix} \end{array}$$

Table 1. The Fundamental Scale

Intensity of Importance	Definition	Explanation
1	Equal Importance	Two activities contribute equally to the objective
2	Weak	
3	Moderate importance	Experience and judgment slightly favor one activity over another
4	Moderate plus	
5	Strong importance	Experience and judgment strongly favor one activity over another
6	Strong plus	
7	Very strong or demonstrated importance	An activity is favored very strongly over another; its dominance demonstrated in practice
8	Very, very strong	
9	Extreme importance	The evidence favoring one activity over another is of the highest possible order of affirmation
Reciprocals of above	If activity *i* has one of the above nonzero numbers assigned to it when compared with activity *j*, then *j* has the reciprocal value when compared with *i*	A reasonable assumption
Rationals	Ratios arising from the scale	If consistency were to be forced by obtaining *n* numerical values to span the matrix

Note that if $a \rightarrow \infty$, , then the priorities converge to the vector (1,0). Thus, we could say that ordinal preferences are a particular case of cardinal preferences.

Let us now consider a group of N voters deciding between candidate A and candidate B. Let $n_{A \succ B}$ represent the number of voters that prefer A to B, and hence, $n_{A \succ B} + n_{B \succ A} = N$. No abstentions are allowed; every voter must rank all the candidates. If we allow for decision makers indifference between candidates, we could assign ½ a point to each candidate without loss of generality, so that $n_{A \succ B} + n_{B \succ A} = N$ is satisfied.

Thus, if we let $v_{A \succ B}$ be the percentage of voters that prefer A to B, $v_{A \succ B} = \frac{n_{A \succ B}}{N}$, then $v_{A \succ B} + v_{B \succ A} = 1$. Note that $\frac{n_{A \succ B}}{n_{B \succ A}} = \frac{v_{A \succ B}}{v_{B \succ A}}$, and the priorities associated with the matrix:

$$\begin{array}{c|cc} & A & B \\ \hline A & 1 & n_{A\succ B}/n_{B\succ A} \\ B & n_{B\succ A}/n_{A\succ B} & 1 \end{array}$$

are given by the normalized to unity values of any column. For candidate A we have:

$$\begin{aligned} \frac{\frac{n_{A\succ B}}{n_{B\succ A}}}{\frac{n_{A\succ B}}{n_{B\succ A}}+1} &= \frac{n_{A\succ B}}{n_{A\succ B}+n_{B\succ A}} = \frac{n_{A\succ B}}{N} = v_{A\succ B} \\ \frac{\frac{n_{B\succ A}}{n_{A\succ B}}}{\frac{n_{B\succ A}}{n_{A\succ B}}+1} &= \frac{n_{B\succ A}}{n_{B\succ A}+n_{A\succ B}} = \frac{n_{B\succ A}}{N} = v_{B\succ A} \end{aligned} \tag{1}$$

Voting requires the counting of votes in favor of each candidate. With intensity of preferences, we cannot add the judgments or use the geometric mean/average to synthesize them. The fact that the judgments could be dispersed (Saaty & Vargas, 2007) or heterogeneous has no bearing on the way we propose to combine them. We need to aggregate them with the condition that when all votes are ordinal, we obtain the result given above in Eq. (1) (Vargas, 2016). Thus, if a voter prefers A to B with intensity a_i, the value of the vote is $\frac{a_i}{a_i+1}$ so that when $a_i \to \infty$, the vote count is 1. Since the priority associated with candidate A is given by $\frac{n_{A\succ B}}{N} = v_{A\succ B}$, then the voting pairwise comparison matrix would be given by:

$$\begin{array}{c|cc} & A & B \\ \hline A & 1 & \sum_{i:A\succ B}\frac{a_i}{1+a_i} \Big/ \sum_{j:B\succ A}\frac{a_j}{1+a_j} \\ B & \sum_{j:B\succ A}\frac{a_j}{1+a_j} \Big/ \sum_{i:A\succ B}\frac{a_i}{1+a_i} & 1 \end{array}.$$

Thus, the priority associated with candidate A is given by:

$$\frac{\sum_{i:A\succ B}\frac{a_i}{1+a_i} \Big/ \sum_{j:B\succ A}\frac{a_j}{1+a_j}}{\left(\sum_{i:A\succ B}\frac{a_i}{1+a_i} \Big/ \sum_{j:B\succ A}\frac{a_j}{1+a_j}\right)+1} = \frac{\sum_{i:A\succ B}\frac{a_i}{a_{i+1}}}{\sum_{i:A\succ B}\frac{a_i}{a_{i+1}} + \sum_{j:B\succ A}\frac{a_j}{a_{j+1}}}. \tag{2}$$

The priority associated with candidate B is obtained in a similar fashion. If all the intensities of preference tend to infinity, we have:

$$\frac{\sum_{i:A\succ B}\frac{a_i}{1+a_i}}{\sum_{i:A\succ B}\frac{a_i}{1+a_i} + \sum_{j:B\succ A}\frac{a_j}{1+a_j}} \underset{\substack{a_i,a_j\to\infty \\ \text{for all } i,j}}{\longrightarrow} \frac{n_{A\succ B}}{n_{A\succ B}+n_{B\succ A}},$$

which represents the percentage of votes candidate A receives.

2.2 Three Candidates

What happens when more than two candidates are available on the ballot, but each voter only votes for one candidate? If we count ballots with ordinal preferences, we can arrange the preferences in a matrix as follows:

$$\begin{array}{c} \\ A \\ B \\ C \end{array} \begin{array}{c} \begin{array}{ccc} A & B & C \end{array} \\ \begin{pmatrix} 1 & {}^{n_{A\succ B}}/_{n_{B\succ A}} & {}^{n_{A\succ C}}/_{n_{C\succ A}} \\ {}^{n_{B\succ A}}/_{n_{A\succ B}} & 1 & {}^{n_{B\succ C}}/_{n_{C\succ B}} \\ {}^{n_{C\succ A}}/_{n_{A\succ C}} & {}^{n_{C\succ B}}/_{n_{B\succ C}} & 1 \end{pmatrix} \end{array}$$

Note that if $\left({}^{n_{A\succ B}}/_{n_{B\succ A}}\right)\left({}^{n_{B\succ C}}/_{n_{C\succ B}}\right) = \left({}^{n_{A\succ C}}/_{n_{C\succ A}}\right)$, the matrix is said to be consistent (Saaty, 1980), and the priorities (or percentage of votes assigned to a candidate) are given by the normalization to unity of any column. However, this is not the case if voters vote for more than one candidate. Thus, the question is: how should we assign percentages to the candidates? Should one only count votes in which a candidate is in the first position? Assume that $\left({}^{n_{A\succ B}}/_{n_{B\succ A}}\right) > 1$, $\left({}^{n_{A\succ C}}/_{n_{C\succ A}}\right) > 1$ and $\left({}^{n_{B\succ C}}/_{n_{C\succ B}}\right) > 1$, and that $\left({}^{n_{A\succ C}}/_{n_{C\succ A}}\right) > \left({}^{n_{B\succ C}}/_{n_{C\succ B}}\right)$, but the consistency condition is not satisfied. It seems clear that A is in first place. In addition, $\left({}^{n_{C\succ A}}/_{n_{A\succ C}}\right) < \left({}^{n_{C\succ B}}/_{n_{B\succ C}}\right)$ and the matrix satisfy *row dominance*; there is an ordering given by the elementwise dominance of the rows. Thus, in this case, candidate A is first, candidate B is second, and candidate C is third. Nonetheless, this positive reciprocal matrix is not consistent.

For 3-by-3 positive reciprocal matrices, any method of deriving priorities from judgments yields the same result (Saaty & Vargas, 1984). The same would be the case if instead of using ${}^{n_{A\succ B}}/_{n_{B\succ A}}$ we were to use intensity of preferences, i.e., $\sum\limits_{i:A\succ B} \frac{a_i}{1+a_i} / \sum\limits_{j:B\succ A} \frac{a_j}{1+a_j}$. So, the fundamental question now is: Would using counts, ${}^{n_{A\succ B}}/_{n_{B\succ A}}$, yield the same or close priorities as using intensity of preferences, $\sum\limits_{i:A\succ B} \frac{a_i}{1+a_i} / \sum\limits_{j:B\succ A} \frac{a_j}{1+a_j}$?

3 Full Rank Voting and Voting with Intensity of Preferences – a Simulation Experiment

To show that full rank voting and voting with intensity of preferences yield very close results, we developed a simulation.

First, we generate random profiles for a number of candidates for n = 3, A, B and C, varying the sample size N from 10, to 100, to 1000, to 10000 and 100000.

A voting profile consists of an ordering of candidates by a group of people. Assume that N = 10 where the group consists of 10 people. We developed an example of a random ranking of 3 candidates as follows:

((3 1 2) (1 3 2) (3 1 2) (1 3 2) (2 3 1) (2 1 3) (1 2 3) (1 3 2) (2 3 1) (3 2 1))

where 1 is the most preferred and 3 is the least preferred in the ordering. Thus, the ordering (3 1 2) means that candidate A is ranked 3rd, candidate B is ranked 1st, and candidate C is ranked 2nd. There are $3! = 6$ different potential rank orders. An example of the random voting profile for the sample generated above is given in Table 2.

Table 2. Randomly generated profile

Number of votes	1	3	1	2	2	1
A	1	1	2	2	3	3
B	2	3	1	3	1	2
C	3	2	3	1	2	1

This profile generates the following pairwise voting matrix:

$$\begin{array}{c} \\ A \\ B \\ C \end{array} \begin{array}{c} \begin{array}{ccc} A & B & C \end{array} \\ \begin{pmatrix} 1 & 6/4 & 1 \\ 4/6 & 1 & 4/6 \\ 1 & 6/4 & 1 \end{pmatrix} \end{array} \begin{array}{c} \text{Priorities} \\ 0.375 \\ 0.25 \\ 0.375 \end{array}$$

The (1,2) entry given by 6/4 represents that 6 people prefer A to B, and 4 people prefer B to A.

Second, we assign intensity of preferences at random from the 1–9 scale while preserving the ordinal preferences. For example, for the ranking (1 2 3) we construct a 3-by-3 reciprocal matrix using random values from the 1–9 scale that yields priorities that preserve the ranking (1 2 3). To ensure that rank is preserved we generate intensities (judgments from the 1–9 scale) satisfying row dominance. Row dominance is a necessary and sufficient condition for rank preservation (Saaty & Vargas, 2012). An example of such a matrix could be given by:

$$\begin{pmatrix} 1 & 3 & 5 \\ 1/3 & 1 & 2 \\ 1/5 & 1/2 & 1 \end{pmatrix} \begin{array}{c} \text{Priorities} \\ \begin{pmatrix} 0.4907 \\ 0.4286 \\ 0.0807 \end{pmatrix} \end{array}$$

We do this for every ranking generated. Even if a ranking appears multiple times, e.g., ranking (1 3 2) appears 3 times, for every occurance we also generate a random set of intensity of preferences.

Third, we combine the judgements as described in Sect. 2. A matrix obtained from the rankings given in Table 2 is shown in Table 3 below.

If instead of having a small group ($N = 10$) we had a larger group, $N = 100$, we perform the three steps described above: (1) generate 100 random rankings of the three candidates, (2) assign intensity of preferences from the 1–9 scale at random satisfying row dominance, and (3) extract the matrix of combined intensity of preferences. The pairwise voting matrix of counts is given in Table 4.

Assigning random intensity of preferences from Saaty's 1–9 scale while preserving the ordinal preferences yields the pairwise reciprocal matrix given in Table 5.

Table 3. Matrix of combined intensity of preferences

	A	B	C	Priorities
A	1.0000	1.5048	0.9094	0.4074
B	0.6646	1.0000	0.5389	0.2301
C	1.0996	1.8557	1.0000	0.3624

Table 4. Pairwise voting matrix for N = 100

Counts	A	B	C	Priorities
A	1	48/52	49/51	0.3201
B	52/48	1	51/49	0.3468
C	51/49	49/51	1	0.3332

Table 5. Pairwise reciprocal matrix corresponding to random intensity of preferences

Int. of Pref.	A	B	C	Priorities
A	1.0000	0.9351	1.3765	0.3297
B	1.0694	1.0000	1.0841	0.3499
C	0.7265	0.9224	1.0000	0.3204

Increasing the size of the voting group to N = 1000, we obtain the following pairwise voting matrix of counts (Table 6) and the corresponding pairwise reciprocal matrix of combined intensities (Table 7).

Table 6. Pairwise voting matrix of counts

Counts	A	B	C	Priorities
A	1	508/492	502/498	0.3378
B	492/508	1	501/499	0.3302
C	498/502	400/501	1	0.3320

For N = 10,000 we have Tables 8 and 9, and for N = 100,000 we have Tables 10 and 11.

For larger groups of candidates, the results are similar. Repeating the three steps in the simulation shown above for 4 candidates, there are 4! = 24 different rankings. For a sample of size N = 10, we obtained the results given in Table 12.

Table 7. Pairwise reciprocal matrix of combined intensity of preferences

Int. of Pref.	A	B	C	Priorities
A	1.0000	1.0464	1.0110	0.3396
B	0.9557	1.0000	0.9557	0.3297
C	0.9891	1.0464	1.0000	0.3307

Table 8. Pairwise voting matrix of counts

Counts	A	B	C	Priorities
A	1	4975/5025	4935/5065	0.3293
B	5025/4975	1	4959/5041	0.3326
C	5065/4935	5041/4959	1	0.3381

Table 9. Pairwise reciprocal matrix of combined intensity of preferences

Int. of Pref.	A	B	C	Priorities
A	1.0000	0.9948	0.9729	0.3297
B	1.0052	1.0000	0.9848	0.3322
C	1.0278	1.0154	1.0000	0.3381

Table 10. Pairwise voting matrix of counts

Counts	A	B	C	Priorities
A	1	50068/49932	49991/50009	0.3336
B	49932/50068	1	49927/50073	0.3327
C	50009/49991	50073/49927	1	0.3337

Table 11. Pairwise reciprocal matrix of combined intensity of preferences

Int. of Pref.	A	B	C	Priorities
A	1.0000	0.9948	0.9729	0.3297
B	1.0052	1.0000	0.9848	0.3322
C	1.0278	1.0154	1.0000	0.3381

Note that the priorities from the pairwise voting matrix (the matrix of counts) and the priorities of the pairwise reciprocal matrix obtained from assigning random intensity of preferences while preserving ordinal preferences are very close.

Table 12. Results for n = 4 candidates and N = 10

Number of Votes	2	1	1	2	1	1	1	1
A	1	2	2	3	3	3	4	4
B	4	3	1	4	1	1	3	3
C	3	1	4	2	4	2	1	2
D	2	4	3	1	2	4	2	1

Counts	A	B	C	D	Priorities
A	1	5/5	4/6	5/5	0.2151
B	5/5	1	3/7	4/6	0.1729
C	6/4	7/3	1	3/7	0.2701
D	5/5	6/4	7/3	1	0.3419

Int. of Pref.	A	B	C	D	Priorities
A	1.0000	1.0372	0.5370	1.0998	0.2115
B	0.9641	1.0000	0.4580	0.8417	0.1837
C	1.8623	2.1834	1.0000	0.4467	0.2854
D	0.9092	1.1880	2.2386	1.0000	0.3194

For a random sample of size N = 1000, we have the following frequencies for the 24 possible rankings:

Frequencies for a random sample N=1000

	46	44	38	51	43	41	36	50	41	49	27	41	43	39	46	50	36	37	41	56	47	33	31	34
	1	1	1	1	1	1	2	2	2	2	2	2	3	3	3	3	3	3	4	4	4	4	4	4
Rank	2	2	3	3	4	4	1	1	3	3	4	4	1	1	2	2	4	4	1	1	2	2	3	3
	3	4	2	4	2	3	3	4	1	4	1	3	2	4	1	4	1	2	2	3	1	3	1	2
	4	3	4	2	3	2	4	3	4	1	3	1	4	2	4	1	2	1	3	2	3	1	2	1

which yields the following pairwise voting matrix of counts:

Counts	A	B	C	D	Priorities
A	1	247/253	66/59	253/247	0.2566
B	253/247	1	269/231	527/473	0.2680
C	59/66	231/269	1	19/21	0.2278
D	247/253	473/527	21/19	1	0.2476

and the estimated pairwise matrix of intensity of preferences:

Int. of Pref.	A	B	C	D	Priorities
A	1	0.9891	1.1270	1.0254	0.2580
B	1.0110	1	1.1807	1.1104	0.2677
C	0.8873	0.8470	1	0.9028	0.2264
D	0.9752	0.9005	1.1076	1	0.2478

As the sample size increases, the frequency distribution of the number of different rankings tends to uniformize around the value N/n! In the example, $1000/24 \approx 41.67$.

There is no loss in generality showing through simulation that even for relatively small samples, the pairwise voting matrices for counts and intensity of preferences, and their corresponding priorities are close. Of course, the pairwise voting matrix using the counts is the correct matrix to use to derive the percentage of votes corresponding to the candidates.

Here is another example in which we assume that not all rankings are equiprobable.

N=48

Number of Votes	2	1	1	20	21	1	1	1
A	1	2	2	3	3	3	4	4
B	4	3	1	4	1	1	3	3
C	3	1	4	2	4	2	1	2
D	2	4	3	1	2	4	2	1

N=48

Counts	A	B	C	D	Priorities
A	1	23/25	24/24	5/43	0.0910
B	23/25	1	23/25	24/24	0.2055
C	24/24	25/23	1	3/45	0.0918
D	43/5	24/24	45/3	1	0.6116

Int. of Pref.	A	B	C	D	Priorities
A	1.0000	0.9116	1.0422	0.1146	0.0942
B	1.0970	1.0000	0.9191	1.0140	0.2097
C	0.9595	1.0880	1.0000	0.0756	0.0954
D	8.7236	0.9862	13.2274	1.0000	0.6006

We have illustrated that full rank voting, even for relatively small number of voters, provides results (candidates' priorities) that are close to the ones obtained with the actual count of the votes.

Let us now explain how rank voting is used today in a political context.

4 How is Rank Voting Used Today

We need to make the distinction between full rank voting and rank-choice voting (RCV) or instant runoff voting. The former makes the voter rank all the candidates, (a vote is not valid without all candidates being ranked), while the latter does not, (some candidates may be left unranked). Thus, with full rank voting the pairwise matrix of counts always has entries of ratios whose denominator are non-zero.

Rank voting lets the voters rank the choices in order of preference. In addition, they can also choose just one candidate, that means that they prefer one candidate over all others, so all other candidates are ranked equally below the preferred one. In a scenario involving more than two candidates, the winning candidate is determined by securing more than 50 percent of the votes. Should no candidate surpass the 50 percent threshold, the candidate with the lowest rank is eliminated. Subsequently, the second-choice preferences of the eliminated candidate's voters are added to the higher-ranked candidates. This iterative process continues until a candidate accumulates the necessary majority, surpassing the 50 percent mark in votes.

For example, assume that we have 48 votes as follows:

N=48								
Number of Votes	2	1	1	20	21	1	1	1
A	1	2	2	3	3	3	4	4
B	4	3	1	4	1	1	3	3
C	3	1	4	2	4	2	1	2
D	2	4	3	1	2	4	2	1

The number of votes the candidates received as first choice are:

N=48									Number of	Percent
Number of Votes	2	1	1	20	21	1	1	1	First ank votes	First rank votes
A	1	2	2	3	3	3	4	4	2	4.17%
B	4	3	1	4	1	1	3	3	23	47.92%
C	3	1	4	2	4	2	1	2	2	4.17%
D	2	4	3	1	2	4	2	1	21	43.75%

N=48									Number of	Percent
Number of Votes	2	1	1	20	21	1	1	1	First ank votes	First rank votes
A	1	2	2	3	3	3	4	4	2	4.17%
B	4	3	1	4	1	1	3	3	23	47.92%
C	3	1	4	2	4	2	1	2	2	4.17%
D	2	4	3	1	2	4	2	1	21	43.75%

Which does not yield a winner since no one received more than 50 percent of the votes. In this case, the candidate with the least number of votes, candidate C, is eliminated and the votes reallocated, by adjusting the ranks of the remaining candidates in the ballots:

N=48									
Number of Votes	2	1	1	20	21	1	1	1	First Ranked vote
A	0	0	1	1	2	1	2	2	3
B	2	1	0	2	0	0	1	1	23
D	1	2	2	0	1	2	0	0	22

Now, A is eliminated, and the first rank votes redistributed:

N=48									Number of
Number of Votes	2	1	1	20	21	1	1	1	First ranked vote
B	1	0	0	1	0	0	1	1	24
D	0	1	1	0	1	1	0	0	24

Which would lead to a runoff election because no candidate received more than 50 percent of the votes. However, if we use the proposed method, full rank voting, in which we not only consider first ranked votes but the results of all paired contexts, A vs. B, A vs. C, ..., D vs. C, we obtain the following pairwise voting matrix:

N=48

Counts	A	B	C	D	Priorities
A	1	23/25	24/24	5/43	0.0910
B	23/25	1	23/25	24/24	0.2055
C	24/24	25/23	1	3/45	0.0918
D	43/5	24/24	45/3	1	0.6116

Int. of Pref.	A	B	C	D	Priorities
A	1.0000	0.9116	1.0422	0.1146	0.0942
B	1.0970	1.0000	0.9191	1.0140	0.2097
C	0.9595	1.0880	1.0000	0.0756	0.0954
D	8.7236	0.9862	13.2274	1.0000	0.6006

which yields D as the winner of the election. Note that D beats all other candidates, except for B. The voters think that D and B are equally preferred in the rank choice vote approach (see the fourth row of the matrix of counts above). Note that the matrix of intensity of preferences generated at random yields very close results.

This raises the question: how should the priorities (or percentage of votes for the candidates) be derived from the rankings available? Why should we only use the vote for the most preferred candidate? Should we not use all the rankings? After all, the rankings are used to reallocate votes when candidates are eliminated. Why not use them from the beginning? Australia uses full rank voting in all parliamentary elections. A ballot is acceptable if all the candidates are ranked. A ballot is unacceptable if one or more of the candidates are not ranked.

5 How Should the Winner of Full Rank Voting Be Computed?

The practice of measuring voting preferences has a historical lineage dating back to at least the 13th century, as noted by (Colomer, 2013). Several of the earliest recorded methods, which continue to be influential and are recognized as a significant division in approaches emerged in France during the 1700s (McLean, 1990). Pioneered by early political scientists and mathematicians such as Jean-Charles, Chevalier de Borda in 1781, and Marie Jean Antoine Nicolas de Caritat, Marquis de Condorcet in 1785. These methods laid the foundation for the contemporary understanding and application of voting systems (Young & Levenglick, 1978).

Let $\mathfrak{A} = \{a_1, ..., a_m\}$ be a set of candidates and $\mathbb{N} = \{1, 2, 3, ...\}$ a set of voters. A *preference order* $\sigma = \{a_{i_1}, ..., a_{i_m}\}$ represent the ranking that the voters assign to the candidates. $L(\mathfrak{A})$ represents all the m! preference orders. A group of voters has the *profile* $M \subsetneq \mathbb{N}$ and may be mapped as $\phi : M \to L(\mathfrak{A})$. All profiles may be represented as Φ. For $\sigma \in L(\mathfrak{A})$ and $\phi \in \Phi$, $n_\sigma(\phi)$ represents the number of those with profile ϕ with preference order σ. $f : \Phi \to L(\mathfrak{A})$ is a *preference function* that maps a set of profiles Φ to the set of ordered candidates. The map from the set of profiles Φ to the set of nonempty subsets of $\mathfrak{A}$ is the *choice function*. The various orderings in profile ϕ is captured by $(\sigma_1(\phi), ..., \sigma_H(\phi))$ and $v_{ij}[\sigma_h(\phi)]$ are the number of voters who prefer i to j in the ordering $\sigma_h(\phi)$ of profile ϕ. $n_{\sigma_h}(\phi)$ are the number of votes in the ordering $\sigma_h(\phi)$

of profile ϕ. $w_i[\sigma_h(\phi)]$ is the weight assigned to the ith alternative in the ordering $\sigma_h(\phi)$ of profile ϕ.

5.1 Full Rank Voting – the Principal Right (PR) Eigenvector Method

Now we show that the vector we need is given by the principal right eigenvector of the matrix $A(\phi) = \left\{ a_{ij}(\phi) \equiv \frac{v_{ij}(\phi)}{v_{ji}(\phi)} \right\}$ (Vargas, 2016).

Consider, for example, the voting profile:

$$\begin{array}{cccc} (4) & (3) & (5) & (2) \end{array}$$
$$B = \begin{pmatrix} a_1 & a_2 & a_1 & a_3 \\ a_2 & a_3 & a_3 & a_2 \\ a_3 & a_1 & a_2 & a_1 \end{pmatrix}$$

For this profile, the pairwise voting matrix and the corresponding principal eigenvector are given by:

$$\begin{array}{cc} \text{Voting Matrix} & \text{PR - Eigenvector} \\ \begin{pmatrix} 1 & 9/5 & 9/5 \\ 5/9 & 1 & 7/7 \\ 5/9 & 7/7 & 1 \end{pmatrix} & \begin{pmatrix} 0.529 \\ 0.412 \\ 0.059 \end{pmatrix} \end{array}$$

and candidate a_1 wins.

Using the traditional rank voting for this example:

$$\begin{array}{cccc} (4) & (3) & (5) & (2) \end{array}$$
$$B = \begin{pmatrix} a_1 & a_2 & a_1 & a_3 \\ a_2 & a_3 & a_3 & a_2 \\ a_3 & a_1 & a_2 & a_1 \end{pmatrix}$$

would yield,

Candidate	*Ranked_First*
a_1	9
a_2	3
a_3	2

The third candidate a_3 is eliminated, and the 2 votes distributed to the second candidate a_2 who is listed as preferred to the first one, yielding:

Candidate	*Ranked_First*
a_1	9
a_2	5
a_3	eliminated

and the first candidate is still the winner. Note that this is the result if all the rankings are used.

Note that if voters vote for only one candidate, the pairwise voting matrix is consistent, and the priorities derived from it coincide with the actual percentages the candidates received. Why should we not use all the rankings that are closely aligned with voters' intensity of preferences?

5.2 Properties of the PR-Eigenvector Method

The properties of the Principal Right Eigenvector Method were more fully developed in (Vargas, 2016) but herein we provide a summary of the approach. We are not able to infer that alternative i beats alternative j from $a_{ij}(\phi) \equiv \frac{v_{ij}(\phi)}{v_{ji}(\phi)} > 1$ if the voting matrix of pairwise comparison ratios does not satisfy *row dominance*. A reciprocal pairwise voting matrix $A(\phi) = \{a_{ij}(\phi)\}$ satisfies *row dominance* if for any two rows i and j, $a_{ih}(\phi) \geq a_{jh}(\phi)$ or $a_{ih}(\phi) \leq a_{jh}(\phi)$, for all h. Hence, a *profile is row dominant* when the corresponding reciprocal pairwise voting matrix is row dominant. Additionally, $v_{ih}(\phi) + v_{hi}(\phi) = v_{jh}(\phi) + v_{hj}(\phi) = N, a_{ih}(\phi) \geq a_{jh}(\phi)$ implies $v_{ih}(\phi) \geq v_{jh}(\phi)$. Row dominance defines a strong order on the set of alternatives.

1. The PR-eigenvector method applied to profiles satisfying row dominance identifies the Condorcet winner.
2. The PR-eigenvector method applied to profiles that satisfy row dominance iss consistent. A voting method f is *consistent* when given two disjoint profiles, ϕ' and ϕ'', it yields the same consensus ordering, $f(\phi') = f(\phi'')$, resulting in the same consensus ordering on the joint profile $\phi = \phi' \cup \phi'', f(\phi) = f(\phi') = f(\phi'')$. The combination of two separate profiles that satisfy row dominance and produce identical orderings for alternatives also satisfies row dominance and results in the same ordering.
3. The PR-Eigenvector method on profiles with row dominance satisfies independence from irrelevant alternatives. A voting method adheres to the principle of independence from irrelevant alternatives (IIA) when the inclusion or removal of an alternative in a profile does not modify the consensus ordering obtained from the original profile.
4. Further, the PR-Eigenvector method on profiles that satisfy row dominance satisfies the independence of *clones* criterion. A clone refers to a candidate that is identical to another in the pool implying that it neither dominates nor is dominated by the other alternatives. Clones do not alter the preferences among alternatives, ensuring that the resulting reciprocal pairwise voting matrix continues to satisfy row dominance. This condition leads to the expected result.

It is worthwhile to point out that because the PR-eigenvector method satisfy all the properties included in Arrow's impossibility theorem (Saaty & Vargas, 2012), full rank voting is not a counterexample to Arrow's theorem.

6 Conclusions

In this paper we put forth a voting method, we call it Full Rank Voting, that captures intensity of preferences in the democratic process when intensity is expressed by Saaty's fundamental scale. Voting is not the same as decision making by group consensus. Voting needs to capture the preferences of everyone in a group. Traditional voting systems often fail to capture the varying degrees of support or opposition individuals have towards different options, leading to an oversimplification of complex issues and a lack of representation of diverse opinions. By incorporating the intensity of preferences into the voting process, we can enhance democratic legitimacy, improve the accuracy of collective will, and facilitate compromise and consensus-building. The method we propose is robust and has very nice mathematical properties such as Condorcet, consistency, and independence from irrelevant alternatives.

We introduced the idea of using Saaty's fundamental scale, which maps words representing intensity to numerical values, to measure the intensity of preferences. This scale allows us to transform qualitative preferences into quantitative values, enabling a more nuanced understanding of individual choices.

We examined the voting process for two candidates and extended it to the case of three candidates. In the case of two candidates, each voter's preference intensity is considered by assigning a numerical value from the fundamental scale. The voting matrix is constructed based on these intensity values, and the priorities of the candidates are determined by normalizing the columns of the matrix. We demonstrated that even with small sample sizes, the pairwise voting matrices for counts and intensity of preferences yield similar results, indicating the effectiveness of full rank voting in capturing the overall preferences of the electorate.

Furthermore, we conducted a simulation experiment to verify the convergence of pairwise voting matrices as the sample size increases. The results showed that as the sample size grows, the matrices obtained from rank voting and intensity of preferences approach close values. This suggests that rank voting provides reliable results that are comparable to the outcomes obtained from the actual count of votes, even for relatively small sample sizes.

Finally, full rank voting, which incorporates the intensity of preferences, offers a promising approach to enhance democracy by providing a more accurate representation of individual choices, improving the legitimacy of outcomes, and enabling compromise and consensus-building. The research findings support the use of full rank voting as an effective method for capturing the intensity of preferences in democratic decision-making processes, and further studies can explore its application in larger-scale elections and real-world scenarios.

Disclosure of Interests. The authors have no competing interests to declare that are relevant to the content of this article.

References

Colomer, J.M.: Ramon Llull: from 'Ars electionis' to social choice theory. Soc. Choice Welfare **40**, 317–328 (2013)

Fechner, G.: Elements of Psychophysics, vol. 1. Holt, Rinehart and Winston (1966)

Herrera-Viedma, E., Martinez, L., Mata, F., Chiclana, F.: A consensus support system model for group decision-making problems with multigranular linguistic preference relations. IEEE Trans. Fuzzy Syst. **13**, 644–658 (2005)

Herrera, F., Herrera-Viedma, E., Chiclana, F.: Linguistic decision analysis: steps for solving decision problems under linguistic information. Fuzzy Sets Syst. **115**, 67–82 (2000)
McLean, I.: The borda and condorcet principles: three medieval applications. Soc. Choice Welfare **7**(2), 99–108 (1990)
Saaty, T.L.: The Analytic Hierarchy Process. McGraw-Hill Publishers (1980)
Saaty, T.L., Vargas, L.G.: Comparison of eigenvalue, logarithmic least squares and least squares methods in estimating ratios. Math. Comput. Model.Comput. Model. **5**, 309–324 (1984)
Saaty, T.L., Vargas, L.G.: Dispersion of group judgments. Math. Comput. Model.Comput. Model. **46**, 918–925 (2007)
Saaty, T.L., Vargas, L.G.: The possibility of group choice: pairwise comparisons and merging functions. Soc. Choice Welfare **38**(3), 481–496 (2012)
Saaty, T.L., Zoffer, H.J., Vargas, L.G., Guiora, A.: Overcoming the Retributive Nature of the Israeli-Palestinian Conflict. Springer, Cham (2022). https://doi.org/10.1007/978-3-030-83958-1
Vargas, L.G.: Voting with intensity of preferences. Int. J. Inf. Technol. Decis. Mak.Decis. Mak. **15**(4), 839–859 (2016)
Young, H.P., Levenglick, A.: A consistent extension of condorcet's election principle. SIAM J. Appl. Math. **35**(2), 285–300 (1978). http://www.jstor.org/stable/2100667
Zadeh, L.A.: The concept of a linguistic variable and its application to approximate reasoning-I. Inf. Sci. **8**, 199–249 (1975)
Zadeh, L.A.: The concept of a linguistic variable and its application to approximate reasoning-II. Inf. Sci. **8**, 301–357 (1975)
Zadeh, L.A.: The concept of a linguistic variable and its application to approximate reasoning-III. Inf. Sci. **9**, 43–80 (1975)
Zhou, S.-M., Chiclana, F., John, R.I., Garibaldi, J.M.: Type-1 OWA operators for aggregating uncertain information with uncertain weights induced by type-2 linguistic quantifiers. Fuzzy Sets Syst. **159**, 3281–3296 (2008)

Author Index

M. Campos Ferreira et al. (Eds.): GDN 2024, LNBIP 509, p. 171, 2024.
https://doi.org/10.1007/978-3-031-59373-4

Printed in the United States
by Baker & Taylor Publisher Services